Philosophy of Mathematics

PRINCETON FOUNDATIONS OF CONTEMPORARY PHILOSOPHY

Scott Soames, *Series Editor*

Philosophical Logic by JOHN P. BURGESS
Philosophy of Language by SCOTT SOAMES
Philosophy of Law by ANDREI MARMOR
Truth by ALEXIS G. BURGESS & JOHN P. BURGESS
Philosophy of Physics: Space and Time by TIM MAUDLIN
Philosophy of Biology by PETER GODFREY-SMITH
Epistemology by ERNEST SOSA
Philosophy of Physics: Quantum Theory by TIM MAUDLIN

PHILOSOPHY OF MATHEMATICS

Øystein Linnebo

PRINCETON UNIVERSITY PRESS
PRINCETON AND OXFORD

Copyright © 2017 by Princeton University Press
Published by Princeton University Press,
41 William Street, Princeton, New Jersey 08540
In the United Kingdom: Princeton University Press,
6 Oxford Street, Woodstock, Oxfordshire OX20 1TR

press.princeton.edu

First paperback printing, 2020
Paperback ISBN 978-0-691-20229-7
Cloth ISBN 978-0-691-16140-2

British Library Cataloging-in-Publication Data is available

This book has been composed in Minion Pro and Archer

Typeset by Nova Techset Pvt Ltd, Bangalore, India

Printed and bound by CPI Group (UK) Ltd, Croydon, CR0 4YY

Contents

Contents

Acknowledgments

I am grateful to many people for discussion of material covered in this book or comments on earlier drafts, in particular Solveig Aasen, Jens Erik Fenstad, Dagfinn Føllesdal, Peter Fritz, Leila Haaparanta, Bob Hale, Mirja Hartimo, Richard Heck, Leon Horsten, Toni Kannisto, Frode Kjosavik, Charles Parsons, Agustín Rayo, Stewart Shapiro, Wilfried Sieg, Hans Robin Solberg, James Studd, Mark van Atten, Crispin Wright, and two anonymous referees. The book has been shaped by courses on the philosophy of mathematics that I taught at the universities of Bristol, London, and Oslo. Thanks to all of my students for their reactions and feedback, which were of great help when designing and eventually writing the book. Most of the writing took place during a period of research leave at the Center for Advanced Study, Oslo, whose support I gratefully acknowledge.

Philosophy of Mathematics

Introduction

MATHEMATICS RAISES A WEALTH of philosophical questions, which have occupied some of the greatest thinkers in history. So when writing this book, some hard choices had to be made.

Let me begin with the aim of the book. Its target audience are advanced undergraduates and graduate students in philosophy, but also mathematicians and others interested in the foundations of one of the most successful, but also most puzzling, human endeavors. For the most part, the book does not presuppose much mathematics. Knowledge of elementary logic, the number systems from the natural numbers up through the reals, and some basic ideas from the calculus will be plenty for all except two late chapters devoted to set theory. While some familiarity with the philosophical mode of thinking will be a clear advantage, I have attempted to explain all relevant philosophical concepts.

I make no attempt to hide my own views concerning what is important and what works. Accordingly, my discussion has some general themes that serve to distinguish it from other introductions to the subject. First, Frege figures prominently in the book, both through his own views and his criticism of other thinkers. While my views often differ from Frege's, I share his fundamental conviction that mathematics is an *autonomous science*. Like other sciences, mathematics uses a meaningful language to express truths, ever more of which are discovered. Yet mathematics differs profoundly from the paradigmatic empirical sciences concerning the nature of its subject matter and the methods it employs. Following Frege, I am critical of any kind of formalism or fictionalism that deprives mathematics of its status as a body of truths, and of any attempt to assimilate mathematics to the empirical sciences. Frege famously defended the objectivity of mathematics. Just as geographers discover continents and oceans, so mathematicians explore numbers and sets. The two kinds of object are equally "real" and are described by equally objective truths.

A second theme of the book is how to understand the objects (such as numbers and sets) that mathematics explores. I pay more attention than is customary to the question of whether mathematical objects can be accepted without fully embracing a so-called platonistic conception of them. So I discuss some less demanding conceptions of mathematical objects. Might these objects be explicable in terms of a network of objective mathematical truths? Or might they be constructed by us? Or might they exist only potentially, not actually?

A final theme concerns mathematical knowledge. This knowledge must be explained in a way that links up with the subject matter of mathematics. It is not just an accident that our mathematical beliefs tend to be true. We would like to know why. What is it about our ways of forming mathematical beliefs which ensures that most of the beliefs correctly represent their subject matter? The answer must draw on an account of mathematical evidence. So what evidence do we have for our mathematical beliefs? A variety of answers have been proposed. Perhaps the evidence is logical or conceptual, or broadly perceptual in character, or of some indirect form that flows from mathematical principles' ability to explain and systematize knowledge already established. My approach to the question of mathematical evidence will be *pluralist* and *gradualist*. That is, one form of evidence need not exclude another. And evidence may come in degrees, such that the elementary parts of mathematics enjoy a higher degree of evidence than the more advanced parts, especially those of a highly set-theoretic character.

Space considerations have forced me to downplay some issues to make room for a proper discussion of the themes just described. There is no systematic discussion of the philosophy of mathematics before Frege's pioneering works of the 1880s and 1890s. I give only the briefest of introductions to Plato's and Kant's views on the subject. Traditional geometry receives little attention. Other important topics receive none. Examples include Wittgenstein on mathematics, explanation in mathematics, the philosophy of mathematical practice, the use of experimental and other nontraditional methods in mathematics, and new developments such as homotopy type theory.[1] The first seven chapters cover topics that tend to be included in any good course in the philosophy of mathematics. The last five chapters discuss more recent developments. These chapters are more specialized and somewhat more demanding, both mathematically and philosophically, but are largely independent of one another (except for Chapter 12, which depends on Chapter 10).

---

[1] Useful introductions to these topics can be found in Rodych (2011), Mancosu (2015, 2008), Baker (2015), and Awodey (2014), respectively.

Mathematics as a Philosophical Challenge

1.1 PROBLEMATIC PLATONISM

Mathematics poses a daunting philosophical challenge, which has been with us ever since the beginning of Western philosophy.

To see why, imagine a community that claims to possess a wonderful kind of knowledge resulting from some discipline practiced there. Community members claim that this knowledge has three distinctive characteristics. First, it is *a priori*, in the sense that it doesn't rely on sense experience or on experimentation. Truths are arrived at by reflection alone, without any sensory observation. Second, the knowledge is concerned with truths that are *necessary*, in the sense that things could not have been otherwise. It is therefore safe to appeal to these truths when reasoning not only about how the world *actually is* but also when reasoning about how it *would have been* had things been otherwise. Third, the knowledge is concerned with objects that are not located in space or time, and that don't participate in causal relationships. Such objects are said to be *abstract*.

In fact, the knowledge that our imagined community claims to posses is rather like the knowledge promised by rational metaphysics, which for centuries professed to deliver insights into the ultimate nature of reality and ourselves, based solely on reason and without any reliance on sense experience. Many people today would dismiss such knowledge claims as incredible. And in fact, science and philosophy have developed in ways that now allow this dismissal to proceed fairly smoothly.

The philosophical challenge posed by mathematics is this. Mathematics seems to deliver knowledge with the three distinctive characteristics that are claimed by our imagined community. "The queen of the sciences"—as Gauss famously called mathematics, usurping a title previously reserved for rational metaphysics—seems to be practiced by means of reflection and proof alone, without any reliance on sense experience or experimentation; and it seems to deliver knowledge of necessary truths that are concerned with abstract things such as numbers, sets, and functions. But in stark contrast to rational metaphysics, mathematics is a paradigm of a solid and successful science. In short, by being so different from the ordinary empirical sciences, mathematics is philosophically puzzling; but simultaneously, it is rock solid.

This challenge obviously requires closer examination. Let us begin with mathematics' strong credentials, before we return, in the sections that follow, to its three apparent characteristics. Mathematics is an extremely successful science, both in its own right and as a tool for the empirical sciences. There is (at least today) widespread agreement among mathematicians about the guiding problems of their field and about the kinds of methods that are permissible when attempting to solve these problems. By using these methods, mathematicians have made, and continue to make, great progress toward solving these guiding problems. Moreover, mathematics plays a pivotal role in many of the empirical sciences. The clearest example is physics, which would be unimaginable without the conceptual resources offered by modern mathematics; but other sciences too, such as biology and economics, are becoming increasingly dependent on mathematics. So a wholesale dismissal of mathematics on the grounds that it is philosophically puzzling would be sheer madness. Such a successful discipline cannot be rejected out of hand but needs to be accommodated within our philosophy in some way or other, albeit perhaps with changes to our pretheoretic conception of how the discipline works. Moreover, unlike rational metaphysics, mathematics permeates our current scientific world view and hence cannot be excised from it.

In sum, our challenge is to explain how we can make room within a broadly scientific world view for a science with features as puzzling as those of mathematics. We shall encounter two lines of response. One is to deny some or all of the distinctive features that appear to set mathematics apart from the ordinary empirical sciences and thus cause philosophical puzzlement. Another line

5

of response is to accept that mathematics is more or less as it appears to be and to explain how this is possible. We shall see that the need for such explanations has profoundly shaped the philosophical outlooks of a number of great thinkers.

1.2 APRIORITY

A good way to approach the seeming apriority of mathematics is to read Plato. And a good place to start is the dialogue *Meno*, where Plato describes a slave boy who has been taught no mathematics but is nevertheless able to discover "out of his own head" an interesting geometrical truth about squares, namely that the square of the diagonal is two times the square of each side. In the dialogue, Socrates asks the slave boy some carefully chosen questions, which prompt the boy to reflect on geometry and discover some simple geometrical truths and eventually reason his way to the mentioned fact about squares.

The story of the slave boy is meant to establish two things. First, that mathematical concepts are innate; that is, they are not acquired but form part of the mind's inborn endowment. And second, that mathematical truths are *a priori* and can be known without relying on experience for one's justification. It may be objected that the slave boy relies on experience in order to understand Socrates' questions. Of course he does! But this experience serves only to *trigger* the process that results in geometrical knowledge and doesn't itself constitute *evidence* for this knowledge.[1]

Suppose Plato is right that we possess innate mathematical concepts and *a priori* mathematical knowledge. How can this be? The usual answer from rationalistically inclined philosophers has

---

[1] Plato offers another argument as well for the innateness of mathematical concepts. For instance, in the *Phaedo*, we find an argument from the following two premises. We possess perfectly precise mathematical concepts, e.g., the concept of a circle. But mathematical concepts are never instantiated in the physical world in a perfectly precise way. Consequently, these concepts cannot be derived from experience but must be innate.

been that our "faculty of reason" is the source of such concepts and knowledge. Until more has been said about this faculty and its workings, however, this answer is little more than a pompous redescription of what we set out to explain. Plato, to his credit, recognizes the need to say more. In the *Meno*, he therefore proposes—or at least entertains—an explanation.

> The soul, then, as being immortal, and having been born again many times, and having seen all things that exist, whether in this world or in the world below, has knowledge of them all. (*Meno*, 81cd)

The envisaged explanation is as follows. The soul must have pre-existed the body. In this disembodied existence, the soul has "seen all things"—including, crucially, the objects with which geometry is concerned—and acquired "knowledge of them all." So when the slave boy—and the rest of us, for that matter—seem to acquire mathematical concepts and knowledge, this is in fact nothing but recollection of concepts and truths that our souls encountered when they existed in a purer, disembodied state and had direct access to the objects of mathematics (as well as to a range of abstract, but perfectly real, "forms" or "ideas" that Plato also postulated).

Of course, this explanation has little appeal today. Plato nevertheless deserves our highest admiration for identifying a deep philosophical problem, namely the seeming apriority of mathematics. The mark of philosophical greatness is as much to identify good questions as it is to answer them. And as we shall see throughout the book, Plato's question has shaped the philosophical debate about mathematics right up until this day.

1.3 NECESSITY

Consider any truth of pure mathematics, say $2 + 2 = 4$. It is part of the traditional Platonistic conception of mathematics that this truth is not accidental—as it is accidental that you are currently reading this book—but that $2 + 2 = 4$ is *necessarily* true, that is, true not only as things actually are, but true no matter how things might have been.

One might worry that this necessity claim is idle philosophical speculation and therefore dispensable. This worry is fueled in part by the philosophical controversy that the notion of necessity has generated and the skepticism it has encountered. The necessity claim has real significance, however, despite these genuine difficulties. Consider the role that mathematics plays in our reasoning. We often reason about scenarios that aren't actual. Were we to build a bridge across this canyon, say, how strong would it have to be to withstand the powerful gusts of wind? Sadly, the previous bridge fell down. Would it have collapsed had its steel girders been twice as thick? This style of reasoning about counterfactual scenarios—or alternative "possible worlds," as philosophers like to call them—is indispensable to our everyday and theoretical deliberations alike. Now, part of the cash value of the claim that the truths of pure mathematics are necessary is that such truths can freely be appealed throughout our reasoning about counterfactual scenarios. Had you not been reading this book, or had some girders been twice as thick, $2 + 2$ would still have been 4. Indeed, the truths of pure mathematics can be trusted even in an investigation of how things would have been in scenarios where the laws of nature are different.

The great German mathematician and philosopher Gottlob Frege (1848–1925), who figures prominently in this book, liked to make a similar point in terms of the "domains" that various kinds of truth "govern." The logical and arithmetical truths are said to govern "the widest domain of all; for to it belongs not only the actual, not only the intuitable, but everything thinkable" (Frege, 1953, §14).[2] Presumably, the domain of "everything thinkable" includes everything that is possible in the sense explained above.

Let us pause to note an immediate but important consequence of the necessity of the truths of pure mathematics. Since such truths can freely be appealed to throughout our counterfactual reasoning, it follows that these truths are counterfactually independent of us humans, and all other intelligent life for that

---

[2] Interestingly, Frege (ibid.) thought the truths of geometry governed the strictly smaller domain of everything "intuitable."

matter. That is, had there been no intelligent life, these truths would still have remained the same. Pure mathematics is in this respect very different from humdrum contingent truths. Had intelligent life never existed, you would obviously not have been reading this book. More interestingly, pure mathematics also contrasts with various social conventions and constructions, with which it is sometimes compared.[3] Had intelligent life never existed, there would have been no laws, contracts, or marriages—yet the mathematical truths would have remained the same. These truths can thus be assumed by us actually existing intelligent agents when we reason about this sad intelligence-free scenario.

1.4 ABSTRACT OBJECTS

The final distinctive feature traditionally attributed to mathematics is a concern with abstract objects. An object is said to be *abstract*, we recall, if it lacks spatiotemporal location and is causally inefficacious; otherwise it is said to be *concrete*. While this distinction may not be entirely sharp, it suffices for our present purposes.[4]

Now, it certainly *seems* that mathematics is concerned with abstract objects. Mathematical texts brim with talk about numbers, sets, functions, and more exotic objects yet, and these objects seem nowhere to be found in space and time.[5] It is useful to "factor" the third feature of mathematics into two distinct claims.

Object realism. There are mathematical objects.
Abstractness. Mathematical objects are abstract.

---

[3] See, e.g., Feferman (2009) and Hersh (1997).
[4] See Rosen (2014) for further discussion.
[5] According to Maddy (1990), sets of concrete objects are located where their elements are located and thus qualify as concrete (cf. §8.4). We shall not take a stand on this. If need be, let us restrict the abstractness claim to "pure" mathematical objects such as numbers and pure sets.

While object realism was endorsed already by Plato, the first clear defense of it was due to Frege. Consider the following sentences:[6]

(1) Evelyn is prim.
(2) Eleven is prime.

The two sentences seem to have the same logical structure, namely a simple predication based on a proper name, which refers to an object, and a predicate, which ascribes some property to this object. As Frege argued, for a sentence of this simple subject-predicate form to be true, the proper name must succeed in referring to an object, and this object must have the property ascribed by the predicate (cf. §2.3). Moreover, (2) *is* true, as anyone who possesses even basic arithmetical competence will confirm. It follows that 'Eleven' must succeed in referring to an object, and hence there are mathematical objects.

Of course, the argument is not beyond reproach. We shall encounter various challenges to it throughout the book. Perhaps the claims of mathematics cannot be taken at face value. Or perhaps they aren't true after all. For now, however, it suffices to observe that the argument has sufficient force to shift the burden of proof onto opponents, who need to explain where they think the argument goes wrong.

The claim that mathematical objects are abstract has been less controversial. It is not hard to see why. If possible, our philosophical account of mathematics should avoid claims that would render our ordinary mathematical practice misguided or inadequate. But if mathematical objects had spatiotemporal location, then our ordinary mathematical practice would be misguided and inadequate. We would then expect mathematicians to take a professional interest in the location of their objects, just as zoologists are interested in the location of animals. By taking mathematical objects to be abstract, our actual practice becomes far more appropriate.

In contemporary philosophy, the word "platonism" (typically with a lowercase 'p') is often used in a more general sense than

[6] The pair of examples is due to Burgess (1999, p. 288).

anything we can ascribe to Plato (and which thus counts as "Platonistic" with an uppercase 'P'). The platonist conception of mathematics does not stop with the claim that there are abstract mathematical objects. It adds a claim about the robust reality of these objects:

> **Reality.** Mathematical objects are at least as real as ordinary physical objects.

Admittedly, this claim is not very precise. Throughout the book, we shall consider some ways to sharpen the claim and assess their plausibility.[7] Plato defended an extremely strong version of the claim. He ascribed to mathematical objects a higher degree of reality or "mode of being" than that of ordinary concrete objects. An imperfect chalk circle on the board is just a pale metaphysical shadow of the perfect abstract circle. The latter is ontologically primary to the former. Few would today follow Plato that far.

There are other interpretations on which Reality is quite plausible, however. One measure of the reality of mathematical objects concerns their independence of intelligent agents and their language, thought, and practices. We have already discussed a counterfactual analysis of this independence (cf. §1.3). Had there been no intelligent life, $2 + 2$ would still have been 4. The independence claim can also be cashed out in terms of a contrast between discovery and invention. It is part of our experience of doing mathematics that mathematical facts are discovered, not invented. Assume you set out to solve some hard mathematical problem and after weeks of hard work finally find the answer. It seems that the answer was already there, waiting for you. The answer was discovered, not made up. Frege therefore compares mathematicians with geographers:

> Just as the geographer does not create a sea when he draws borderlines and says: the part of the water surface bordered by these lines I will call Yellow Sea, so too the mathematician cannot properly create anything by his definitions. (2013, I, xiii)

---

[7] Some challenges to Reality that hold on to object realism will be discussed in §§2.5, 4.4, and 5.2.

The point is even enshrined in U.S. patent law, which permits one to patent inventions but not mathematical truths or laws of nature.

Summing up, we shall follow recent philosophical practice and use the label "platonism" (with a lowercase 'p') for the conjunction of the three claims discussed in this section. As we have seen, the traditional conception of mathematics inherited from Plato includes additional claims about apriority and necessity, which are not included in our definition of 'platonism'.

1.5 THE INTEGRATION CHALLENGE

Assume we take mathematical language and practice more or less at face value and accept some version of the platonistic conception. A philosophical challenge arises. Can we make sense of a science that works in this way? Can we explain how human beings in a seemingly *a priori* way acquire knowledge of necessary truths concerned with abstract objects? As we have seen, this challenge has been with us since Plato's *Meno*.

Let me explain how the challenge is best developed.[8] As Quine (1960) emphasized, our theorizing always begins *in medias res*, that is, in the context of science as we find it. And science, as we find it, obviously includes mathematics. The challenge is to use science, as we find it, to try to understand our practice of mathematics. Just as science can be used to study navigation in birds and primates' knowledge of their environment, it can also be used to investigate human knowledge of mathematics.

Our investigation has two parts. First, we need to get clear on the subject matter of mathematics. What is mathematics about? Is it really concerned with abstract objects, as the platonistic conception would have it? To answer these questions, we obviously need to listen to what mathematics itself has to say

---

[8] Recent discussions of the challenge often focus on the version developed in Benacerraf (1973), which requires a causal connection between the knower and the known. I find this focus unfortunate, for reasons that surface below but are more fully explained in §7.1.

about numbers, sets, and everything else that it studies. As Frege emphasized, however, the questions are also in part concerned with language. How should the language of mathematics be analyzed? Should apparent talk about numbers and sets be taken at face value? This concern with language means that we shall also need assistance from linguistics and perhaps also psychology. Second, we need to understand how mathematicians and others with some degree of mathematical competence arrive at their mathematical beliefs. How do mathematicians settle on their first principles (or *axioms*), and how do they use these to prove mathematical results (or *theorems*)? In this investigation, the psychology and history of mathematics will clearly be relevant.

The challenge is to make our answers to these two sets of questions mesh. How is it that our ways of forming mathematical beliefs are responsive to what mathematics is about? How are the practices and mechanisms by which we arrive at our mathematical beliefs conducive to finding out about whatever reality mathematics describes? In short, why is it not just a happy accident that our mathematical beliefs tend to be true? There must be something about what we do that keeps us on the right track. Since the challenge is to integrate the metaphysics of mathematics (namely, what mathematics is about) with its epistemology (namely, how we form our mathematical beliefs), we shall call this *the integration challenge*.[9]

Since the questions we are asking are hard and the literature abounds with misapprehensions, I would like to warn against three ways in which the challenge can be misunderstood.[10] Clearly, the challenge goes beyond any individual branch of science, and in this sense, it is distinctively philosophical. But it would be a mistake to think that the challenge is external to science as a whole, as for instance Descartes' "first philosophy" famously aspired to be. The challenge arises within science, broadly construed, when we decide to investigate not birds'

---

[9] This label is due to Peacocke (1999).
[10] A fuller discussion can be found in Linnebo (2006).

navigation or primates' knowledge of their environment but our own knowledge of mathematics.

Next, it would be wrong to think that the challenge is prejudiced against mathematics. True, the challenge would have been prejudiced if we had insisted on a *causal connection* between the subject matter of mathematics and mathematicians' beliefs: for such a connection might be inappropriate in mathematics. But there is no such insistence. The nature of the connection is left wide open. All we want is to know why it is not an accident that our mathematical beliefs tend to accurately represent what they are about. Moreover, we must bear in mind that the challenge arises *in medias res* and that mathematics is an important part of science, as we find it. We are therefore free to use as much mathematics as we please when trying to answer the challenge.

Finally, one might think the challenge is trivial, given that we are allowed to use mathematics when trying to answer it. Isn't it obvious that mathematics will judge itself to be in good order? The case seems much like Wittgenstein's example of a man who buys a second copy of a newspaper in order to confirm what the first copy says! This reaction would be mistaken, however. To see why, a comparison is useful. Assume we are interested in explaining the nature and role of perception in our belief formation. We can give an account of how perceptual beliefs are reliable by explaining how light is reflected from surfaces, impinges on our retinas, thus triggering nerve endings, and so on. Of course, this account of the reliability of perception is itself reliant upon perceptual knowledge: otherwise we could not appeal to light, surfaces, and retinas. But this mild circularity is unproblematic. We are proceeding *in medias res*, not trying to convince a skeptic. The mild circularity does therefore not trivialize our project. We are presupposing *that* our perceptual beliefs are reliable in order to explain *why* they are reliable. So there is some distance between what we are presupposing and what we are trying to explain. This distance means that success is not guaranteed. We are not just buying a second copy of the same newspaper. The same goes for our investigation of the formation of mathematical knowledge. By allowing the investigation to use mathematics, we are presupposing *that* our mathematical beliefs are reliable, at least for the most part. But this does not guarantee success in our task of explaining *why* these beliefs are reliable.

In sum, the integration challenge is both legitimate (contrary to the first two misapprehensions) and substantive (contrary to the third).

1.6 KANT'S VIEW ON MATHEMATICS

How might the philosophical challenge posed by mathematics be met? Let us for now restrict ourselves to the vast class of views that take us to possess mathematical knowledge. (We shall return to some views that deny this.) The obvious question is then: what sort of knowledge is this? Until fairly recently, answers to this question tended to be given in terms of Kant's distinctions between *a priori* and *a posteriori*, and analytic and synthetic. Although hardly unproblematic, this classification provides a shared frame of reference, which many of the thinkers we shall study use in order to locate their own views and those of their interlocutors.

The basic idea of the distinction between *a priori* and *a posteriori*, we recall, is that a justification or piece of knowledge is *a priori* if it is independent of experience, and otherwise *a posteriori*. Some hard questions immediately arise. What is the relevant class of experiences? While sense perception clearly conflicts with apriority, the experience that accompanies the discovery of a mathematical proof does not. Furthermore, what is the relevant notion of "independence" of experience? As we saw, even Meno's slave boy needs the experience of hearing Socrates' questions in order to trigger the process of mathematical learning. Kant is aware of this complication and admits that "all cognition commences *with* experience." What he denies is that all cognition "arise[s] *from* experience" (Kant, 1997, B1). Instead he holds that some of our knowledge has its "source" in our own cognitive faculties, in ways to which we shall return.

Kant defines a judgment as *analytic* if "the predicate *B* belongs to the subject *A* as something that is (covertly) contained in this concept *A*" and *synthetic* otherwise (ibid., A6/B10). All analytic judgments are thus *a priori*, since this conceptual containment can be established without any substantive reliance on sensory experience. Analytic judgments are also said to be "explicative," while synthetic ones are "ampliative." For example, Kant holds that "Bodies are extended" is analytic, whereas "Bodies are heavy" is synthetic. But it is a problem that his definition works only for simple subject-predicate judgments of the form "*A* is *B*." How should the definition be extended to the language of mathematics, whose statements are typically far more complex? One option is to rely on another characterization that Kant gives of the analytic truths, namely as those that are based on the principle of contradiction (ibid., A151/B191). This characterization suggests that all logical truths may qualify as analytic. This idea is later adopted by Frege (cf. §2.2).

With this two-by-two classification in place, let me provide a brief sketch of Kant's own view. In traditional fashion, Kant insists that mathematical knowledge is *a priori*. The truly novel part of his view is that, despite being *a priori*, mathematical knowledge is not analytic; in other words, that mathematical knowledge is synthetic *a priori*. According to Kant, this form of knowledge had not previously been identified and was not at all understood. To say that much of Kant's theoretical philosophy was shaped by his desire to account for our possession of synthetic *a priori* knowledge is no exaggeration.

Why does Kant hold that mathematical knowledge is synthetic? He discusses examples from arithmetic as well as geometry. Consider the judgment that $7 + 5 = 12$. According to Kant, "[t]he concept of twelve is by no means already thought merely by my thinking of that unification of seven and five" (ibid., B15). Rather, in order to establish that this judgment is true, we must go beyond the concepts involved and bring in the aid of intuitions to represent these concepts, for example, by producing the relevant numbers of fingers or points. So arithmetical truths are not grounded in facts about conceptual containment but are "ampliative" and thus synthetic. The case of geometry is analogous. To establish that the shortest line between two points is straight, it is futile to contemplate the concepts involved in this truth. Instead we need to bring in intuition to draw—perhaps in "pure imagination"—the shortest line between two points. We can then perceive that this line is straight.

One of the central philosophical questions, according to Kant, is how the new form of knowledge that he had identified— the synthetic *a priori*—is possible. In particular, how can we make sense of a science such as mathematics? In order to answer these questions Kant finds it necessary to undertake his famous "Copernican turn," which ushers in his "transcendental idealism." Let me explain. On the ordinary, "pre-Copernican" conception of knowledge, objects have their properties independently of us, and our mental representations must conform to these objects in order to count as knowledge. But according to Kant, this conception is unable to accommodate synthetic *a priori* knowledge. Since the knowledge is synthetic, it is about objects; but if our representations must conform to their objects, then these objects must somehow affect us, thus rendering the knowledge *a posteriori*. The only solution is to take the Copernican turn:

> If intuition has to conform to the constitution of the objects, then I do not see how we can know anything of them *a priori*; but if the object ... conforms to the constitution of our faculty of intuition, then I can very well represent this possibility to myself. (Ibid., Bxvii)

That is, we need to reverse the usual order of epistemic conformity. It is only when we recognize that the world must conform to "the constitution of our faculty of intuition" that we become able to explain how arithmetical and geometrical truths can be simultaneously synthetic and *a priori*. The claim is thus that, in order to explain how mathematics is possible, we need to adopt Kant's transcendental idealism, which results from this reversed order of conformity. Convincing or not, this is indisputably one of the most dramatic episodes in Western thinking about mathematics, which rivals Plato's discussion in importance and influence.

1.7 An Overview of the Options

A brief overview of the accounts of mathematics to be considered in the book may be useful. Let us begin with views that assume mathematical knowledge. What kind of knowledge is this? Kant's classification enables three possible answers.

One answer is that mathematical theorems are synthetic *a priori*. This view can be combined with a "pre-Copernican" metaphysics, as exemplified by Plato (to describe his view in patently anachronistic terms), or with a "Copernican" metaphysics, as exemplified by Kant himself and more recently the intuitionism of L.E.J. Brouwer.

A second answer is that the theorems of mathematics are *a posteriori* (and hence also synthetic, since all analytic truths are *a priori*). As we have seen, this view contradicts a long tradition of taking the epistemology of mathematics to be fundamentally different from that of the ordinary empirical sciences. This has not prevented adventurous thinkers from exploring and defending the view. An early example is John Stuart Mill, and a more recent one, W. V. Quine. As we shall see, however, Quine's preferred option is to abandon Kant's distinction between analytic and synthetic.

The final option is that mathematical truths are analytic. Kant regarded this option as hopeless. The concept of 12 is simply not contained in the concepts of 7, 5, or addition. Kant may well be right that mathematical truths aren't analytic—*according to his own definition*. But as noted, Kant's official definition works only for subject-predicate statements. These simple statements figure centrally in the Aristotelian logic that still dominated at Kant's time and that Kant seems to have regarded as definitive (1997, Bviii). The next dramatic episode in our story concerns Frege, who is the father of the modern quantificational logic that eventually superseded Aristotle's. Frege's new logic recognizes a wealth of statements whose logical structure is far more complex than those to which Kant's official definition applies. This opens up a promising new option. Perhaps some of these statements should be seen as analytic, although they don't qualify as such by Kant's definition. Perhaps the impoverished logic available at

Kant's time resulted in a dramatic underestimation of the sphere of analytic knowledge. This is precisely what Frege thought. He pioneered the influential view known as *logicism*, which holds that at least all truths about numbers—the naturals up through the reals—are reducible to logic and thus analytic.

The following table summarizes our discussion so far.

	analytic	synthetic
a priori	Frege	Plato (pre-Copernican)
		Kant, Brouwer (Copernican)
a posteriori	✗	Mill, Quine

As mentioned, some views fall outside of this table because they deny that mathematical theorems are true. The denial is not quite as desperate as it may seem. No one is claiming that all mathematical theorems are *incorrect*. The idea is rather that mathematics, unlike most other sciences, operates with a standard of correctness that is less demanding than truth. For instance, a move in a game such as chess can be correct, although it would be inappropriate to call it true. Indeed, many *formalists* compare mathematics to a game. They deny that mathematical sentences are meaningful and propose instead to understand mathematics as the activity of proving pure formal theorems from purely formal axioms. Moreover, some *fictionalists* view mathematics as a useful fiction, where the standard of correctness isn't literal truth but truth according to some fiction.

1.8 SELECTED FURTHER READINGS

Shapiro (2000, esp. chaps. 1–4) provides a good general introduction to philosophy of mathematics; Horsten (2016) is a good shorter alternative. Plato's *Meno*, especially 80a–86c, and Kant (1997), especially the B-introduction §§I–V and B740–752, are important historical texts. Parsons (1982) and Friedman (1985) are influential discussions of Kant's view of arithmetic and geometry, respectively.

Finally, let me mention some works that are relevant to the book as a whole. Benacerraf and Putnam (1983) is an indispensable collection of classic readings in the philosophy of mathematics. Shapiro (2005b) contains valuable surveys of all the main approaches to the subject. Surveys of a huge variety of philosophical views, concepts, and traditions can also be found in the excellent online *Stanford Encyclopedia of Philosophy.*

Frege's Logicism

FREGE'S PHILOSOPHY OF MATHEMATICS combines two tenets. On the one hand, he was a platonist, who believed that abstract mathematical objects exist independently of us. On the other hand, he was a logicist, who took arithmetic to be reducible to logic. This combination of tenets may seem surprising. It would certainly have been unheard of to Kant, who insisted that objects are "given to us" only through perception or intuition, never by logic or reason alone. The combination also clashes with today's dominant conception of logic, which requires that logical truths be true in all models, including ones devoid of any mathematical objects. It follows immediately that the existence of mathematical objects can never be a matter of logic alone.

In another respect, however, Frege's combination of tenets should seem quite natural. Although concerned with mathematical objects, mathematical truths are extremely general in their applicability. For example, things of absolutely any kind—whether physical, psychological, or abstract—can be counted; and the laws of arithmetic remain valid in any scenario that is thinkable at all. As Frege liked to emphasize, the truths of arithmetic and logic have in common that they govern "the widest domain of all" (1953, §14).

2.1 RIGOR AND FORMALIZATION

It is useful to consider Frege's emphasis on the extreme generality of the truths of arithmetic against the background of mathematical developments in the nineteenth century. Ever since Euclid (ca. 300 BCE), mathematics has relied heavily on *the axiomatic method*. Some particularly self-evident or fundamental truths are identified and given the status as *axioms*, that is, as first principles that need no proof but to which we can freely appeal in all proofs of derived principles, known as *theorems*. This method has proved immensely successful and is now an essential part of mathematical practice.

In light of the axiomatic method, the epistemology of mathematics splits into one question about our knowledge of the axioms and another question about our entitlement to the deductive reasoning used to derive the theorems. The former question is at the heart of the philosophy of mathematics. Before the nineteenth century, axioms were often justified merely on the grounds that they are intuitively obvious. This intuitive obviousness was often a matter of some geometrical construction or observation. A good example is the *intermediate value theorem*, which says a continuous function that for some argument has a value less than some number c and for another argument has a value greater than c must for some argument have value exactly c. This important principle of mathematical analysis used to be regarded as intuitively obvious and requiring no proof. Surely, any continuous function that begins on one side of some horizontal line and ends up on the other side must at some point intersect the line.

In 1817, the great Austrian mathematician, philosopher, and theologian Bernard Bolzano (1781–1848) published what he billed as a "purely analytical proof" of the intermediate value theorem. Instead of regarding the theorem as basic and adequately justified by an appeal to geometrical intuition, Bolzano offered a logical analysis of continuity—much like today's epsilon-delta analysis—and used this analysis to produce a proof of the theorem from precise logical assumptions about the structure of the real numbers.[1] This pioneering work was motivated not by skepticism about geometrical intuition but by a methodological desire to identify the most fundamental

---

[1] On the epsilon-delta analysis, a function f is said to be continuous at an argument a iff, loosely speaking, the value of f for arguments around a can be made arbitrarily close to $f(a)$ by ensuring that the argument is sufficiently close to a; or, precisely speaking, iff for any $\epsilon > 0$ there is a $\delta > 0$ such that $|x - a| < \delta$ ensures $|f(x) - f(a)| < \epsilon$. (As customary, 'iff' is short for 'if, and only if.')

truths from which the theorem follows. The extant justification could at best give the theorem a geometrical foundation, thus limiting its range of validity to that of geometry. As Bolzano realized, however, the theorem has a far greater range of validity. Continuous functions are not the sole property of geometry but can be found wherever one quantity depends on another.

Unfortunately, Bolzano's progressive political and theological views led to trouble with the authorities, severely curtailing his ability to publish and thus limiting his influence. But other nineteenth-century mathematicians (notably Cauchy and Weierstrass) proceeded to complete the program of methodological reform that Bolzano had inaugurated. *The rigorization of analysis*, as the program is known, resulted in precise logical definitions of the central concepts of continuity and limits, as well as the now-standard constructions of the systems of rational, real, and complex numbers. Just as Bolzano had envisaged, the effect was a gradual elimination of appeals to geometrical intuition in favor of far more abstract logical and analytical methods. This new and more abstract foundation ensured a far greater range of validity of the mentioned number systems and the mathematical principles that govern them.

Frege's academic career started at the height of this process of rigorization, and he contributed to the project in two separate ways. The first contribution, published in 1879, concerns the reasoning that takes us from axioms to theorems. Until Frege, this reasoning was largely informal and conducted in natural languages (often German or French), augmented with mathematical symbols. Frege found this informality unacceptable. Even if our axioms are "purely analytic," how do we know that our proofs haven't tacitly relied on intuitive principles, thus compromising the "purity" of our theorems and misleading us about the range of their validity? "So that nothing intuitive could intrude here unnoticed," Frege wrote, "everything had to depend on the chain of inferences being free of gaps" (1879, p. iii). Where there are gaps, intuitive assumptions may intrude. Frege found it impossible to enforce this requirement of gap-freeness in an informal natural language. So he invented the first ever formal language, his *Begriffsschrift*, or "conceptual writing." In this language, he   23 proceeded to lay down precise logical axioms and inference rules, which take us from axioms or theorems already established to further theorems. In short, Frege's aim of gap-free proofs led him to formulate what is now known as a *formal system*, which is an artificial language with clearly formulated axioms and inference rules, all described with mathematical precision.[2] This is arguably the greatest contribution to the axiomatic method since Euclid. As we shall see, the notion of a formal system is also essential to much subsequent theorizing about mathematics, both philosophical and metamathematical.

The logic that Frege articulated is far stronger than the broadly Aristotelian logic used by Kant. Frege identified—and laid down logical principles governing—all the truth-functional connectives and the quantifiers.

In fact, Frege recognized several different kinds of quantifiers, which effect different kinds of generalization. *First-order quantifiers* generalize into the position occupied by a singular term or proper name, while the more controversial *second-order quantifiers* generalize into the position occupied by a predicate. Consider, for example, the claim that Socrates is mortal, symbolized as Ms. While first-order logic allows us to infer that there is someone who is mortal, $\exists x\, Mx$, second-order logic allows us to infer that there is some "concept" (as Frege put it) under which Socrates falls, $\exists F\, Fs$.[3]

2.2 ANALYTICITY EXTENDED

Frege's second contribution concerns the axioms, rather than the reasoning from axioms to theorems. All of the earlier constructions of number systems took for granted the sequence of natural number, namely 0, 1, 2, ... (This prompted Kronecker's famous quip that "God made the integers; all the rest is the work of man.") But what are these numbers, and what is the nature

---

[2] Unlike the later formalists, however, Frege took this language to be meaningful. Cf. §3.1.

[3] See Shapiro (2005a) for an introduction to higher-order logic.

of our knowledge of them? These questions set the agenda for Frege's next major work, *Foundations of Arithmetic*. Motivated by his conviction that the principles of arithmetic are just as general as those of logic, Frege seeks to establish that these principles are analytic and thus to overturn the influential Kantian view that they are synthetic *a priori*.

For this to be so much as an option, Frege needs to extend the Kantian definition of analyticity. He begins by emphasizing the role of proof for the classification of knowledge. The classification of some piece of knowledge as analytic or synthetic, *a priori* or *a posteriori*, concerns not "the psychological, physiological and physical conditions" that made it possible for us to grasp the relevant proposition but rather "the ultimate ground on which the justification for holding it to be true rests" (Frege, 1953, §3). As Bolzano had emphasized, we need to find the best possible proof, or justification, of the proposition and to examine the nature of the assumptions on which this proof rests. If the proof relies only on "general logical laws and definitions," then the proposition is analytic. But if the proof requires assumptions that "belong instead to the domain of a particular science, then the proposition is synthetic" (ibid.). Thus, in order to demonstrate that arithmetic is analytic, Frege needs to provide proofs of its principles that rely only on "general logical laws and definitions" (ibid.).

It may be objected that Frege stacks the cards against Kant by adopting a more inclusive definition of analyticity. Frege anticipates the objection and insists that he has merely "stated accurately" what Kant had in mind (ibid., §3, fn. 1). There is some basis for the response. Kant sometimes characterized the analytic truths as those that are based on the principle of contradiction.[4] So when we realize that logic extends beyond this principle and the Aristotelian logic associated with it, this extended logic should be allowed to ground further analytic truths.

A more serious complaint, to which we shall return, is that Frege has not told us nearly enough about the nature and extent of his "general logical laws."

---

[4] See, e.g., Kant (1997, A151/B191); cf. §1.6.

2.3 THE ANALYSIS OF NUMBER ASCRIPTIONS

How can we prove the principles of arithmetic on the basis of "general logical laws and definitions"? As we have seen, Kant denied that the concept of 12 is "contained in" the concepts of 7, 5, or addition. To do better, we need an analysis of the language of arithmetic that is more discerning than Kant's. Only then will we have a chance to establish that the truths of arithmetic are analytic.

A natural place to start is with *number ascriptions*, that is, statements of the form "the number of *F*s is so-and-so." Frege recognized the importance of the concept *F* to such number ascriptions. "[I]n looking at the same external phenomenon," he observes, "I can say with equal truth 'This is a copse' and 'These are five trees,' or 'Here are four companies' and 'Here are 500 men'" (1953, §46). What changes in these examples is not the "external phenomenon" but the concept—as Frege calls the objective meaning of a predicate—that specifies what we are counting. "This suggests ... that a statement of number contains an assertion about a concept" (ibid.). When I assert "These are five trees," for example, I am saying of the first-level concept TREE that it is quintuply instantiated. Frege is particularly pleased that "[t]he extensive applicability of number can now be explained" as follows. Numbers are ascribed to concepts, which can apply to *absolutely all kinds of object*, "the physical and mental, the spatial and temporal, the non-spatial and non-temporal" (ibid., §48).

The observation that numbers apply to concepts goes only part of the way toward an analysis of number ascriptions. One natural continuation would be to regard numbers as *second-level concepts* that ascribe cardinality properties to first-level concepts. For instance, "These are five trees" might be ana-lyzed as "$5x(\text{TREE}(x))$," much like "There is a tree" is analyzed as "$\exists x(\text{TREE}(x))$." Frege subjects this analysis to a barrage of objections. The most serious is that "[e]very individual number is an independent object" and thus not a second-level concept. As evidence, he points to the fact that we speak of "the number 1," where the definite article serves to class it as an object. In arithmetic this self-subsistence comes out at every turn, as for example in the identity $1 + 1 = 2$. (Ibid., §57)

This argument has attracted much scholarly attention because of its unusual mix of linguistic and metaphysical considerations.[5] The premises involve *linguistic* considerations about how we talk about numbers. But the desired conclusion is *metaphysical*, namely that numbers are objects, not second-level concepts. That is, the conclusion is an instance of what we earlier called "object realism."

In fact, we have already outlined how the argument is meant to work (cf. §1.4). To spell things out further, we need to talk about *semantics*, which is the branch of linguistics concerned with meaning, reference, and truth—in short, with the relation between language and the world. *The Fregean argument*, as I shall call it, relies on the following two premises:

Classical Semantics. The singular terms of the language of mathematics are supposed to refer to mathematical objects, and its first-order quantifiers, to range over such objects.

Mathematical Truth. Most sentences accepted as mathematical theorems are true.

We now reason as follows. Consider sentences that are accepted as mathematical theorems and that contain one or more mathematical singular terms. By Mathematical Truth, most of these sentences are true. Let S be one such sentence. By Classical Semantics, the truth of S requires that its singular terms succeed in referring to mathematical objects. Hence there are mathematical objects.

In fact, the Fregean argument is far more general than Frege's preferred development of it. Frege's own strategy was to defend Mathematical Truth by showing mathematical theorems to be

---

[5] Indeed, according to Dummett (1981, p. 14; 1991, pp. 111–12), this argument initiated "the linguistic turn" in philosophy, which in turn marked the birth of analytic philosophy.

logical truths. As we shall see, however, alternative defenses of this premise are possible as well.

2.4 HOW DO WE GRASP MATHEMATICAL OBJECTS?

Frege realizes that his conclusion that numbers are objects will strike many readers as puzzling. So he goes on to consider some questions. Some of these are easily answered. For example, the question "Where is the number 4?" is answered simply by observing that "[n]ot every objective object [objectives Gegenstand] has a location" (1953, §61). After all, numbers are abstract, not concrete. But this answer prompts another, harder question: "How, then, are numbers to be given to us, if we cannot have any ideas or intuitions of them?" (§62). This is a version of what we have called "the integration challenge" (cf. §1.5).[6] Since numbers cannot be perceived or tracked by means of instruments, how do we manage to refer to them, let alone gain knowledge of them?

Frege's response to this challenge takes us right to the heart of his philosophy:

> Only in the context of a sentence do words have meaning. We must, therefore, define the sense of a sentence in which a number-word occurs. (Ibid., §62)

This requires some unpacking. First, Frege states that the meanings of words need to be explained in the context of complete sentences. Only complete sentences make claims capable of being true or false, and the meaning of a word is a matter of its potential to contribute to the expression of such claims. This has become known as the *context principle*. Next, Frege applies the context principle to the question of how the numbers are "given to us." This is a semantic question about how number terms come to refer to numbers. By the context principle, this question cannot

---

[6] These concerns remained high on Frege's agenda. For example, in Frege (2013, II, §147) we read: "If there are logical objects at all—and the objects of arithmetic are such—then there must also be a means to grasp them, to recognise them."

be answered in isolation, outside of any sentential context. The question must instead be answered by explaining how sentences containing number terms come to be meaningful. In this way, Frege hopes, the context principle will transform the question about our semantic and epistemic "access" to numbers into an easier question about the meanings of arithmetical sentences.

Arithmetical identities are particularly important, Frege thinks. So our first task is to explain their meanings. Frege's explanation relies on his view that numbers are ascribed to concepts. This makes identities of the form "the number of Fs = the number of Gs"—or in symbols, "$\#x\,Fx = \#x\,Gx$"—especially important.[7] Putting things together, the original problem of explaining how the numbers are "given to us" is thus transformed into the problem of explaining the meanings of identities of the form "$\#x\,Fx = \#x\,Gx$." The meanings of other sentences involving number terms can be explained later.

Frege proposes a brilliant analysis of the mentioned identities. He proposes that the meaning of "$\#x\,Fx = \#x\,Gx$" be "recarved" as the claim that the Fs and the Gs can be one-to-one correlated. What it means for the number of knives on a table to be identical with the number of forks, for example, is that each knife is correlated with a unique fork and vice versa. This analysis is attractive. It seems to avoid all use of the number terms whose meaning we are trying to explain. All that the analysis requires are the resources to express that the Fs and the Gs can be one-to-one correlated. And these resources are available in pure second-order logic, where a second-order quantifier can be used to state that there is a relation that one-to-one correlates the Fs and the Gs. Thus, Frege's proposed analysis seems both noncircular and purely logical.

The proposed analysis is encapsulated in the principle

(HP) $\qquad\qquad \#x\,Fx = \#x\,Gx \leftrightarrow F \approx G$

---

[7] To be precise about the syntax: the operator '$\#x$' applied to any formula φ, binds any free occurrence of 'x' in φ, thus producing the singular term '$\#x\,\varphi$' (which may be read as "the number of x's such that φ").

where '$F \approx G$' abbreviates the mentioned claim about one-to-one correlations, or, as we shall put it, that the Fs and the Gs are *equinumerous*. Because Frege claims to be taking a cue from Hume, this has become known as *Hume's Principle*.[8]

Principles such as Hume's are known as *abstraction principles*. This terminology requires some explanation. According to the philosophical tradition, to abstract is to "extract" from a class of things a feature that these things have in common when they are equivalent in some respect. For instance, we abstract the color *red* from a collection of things that are chromatically equivalent. Hume's Principle lends itself to an analogous interpretation. One way in which concepts can be equivalent is by being equinumerous. When we abstract on concepts that are equivalent in this respect, we obtain their cardinality or number. There are many other abstraction principles as well. A favorite example of Frege's describes how directions can be abstracted from lines:[9]

$$(\text{Dir}) \qquad d(l_1) = d(l_2) \leftrightarrow l_1 \parallel l_2$$

That is, a direction is the feature that two lines have in common just in case they are parallel.

How should we understand the "extraction" of a common feature that is shared by all the equivalent things? One traditional account is concerned with the psychology of concept formation. By perceiving many lines and observing that they are parallel to one another, we "extract" the concept of their direction. Frege is hostile to this psychological account, however, and tries to articulate a purely logical alternative, based on his idea of "recarving of meaning." Consider the abstraction principle (Dir). Each instance of its right-hand side has an antecedently available meaning. This meaning is "carved up" in a new way by the corresponding instance of the left-hand side, which refers to a

---

[8] It would have been more appropriate for Frege to cite his contemporary Georg Cantor, who used the principle in his groundbreaking and profound analysis of infinite (cardinal) numbers; cf. §4.2.

[9] Admittedly, we would obtain a better fit with our ordinary concept of direction by considering instead *directed* lines or line segments and the equivalence relation of "co-orientation," defined as parallelism *plus sameness of orientation*. But since all philosophical considerations remain unchanged, we shall not bother.

new kind of object—namely directions—that are not mentioned on the right-hand side. Since this logical account of abstraction is novel, Frege proceeds to discuss some objections. First, he reminds us that the "recarving of meaning" does not assign a new sense to the identity symbol but uses it in its familiar sense. But as Frege observes, this sets him up for a second objection. Doesn't the "recarving" run the risk of violating the familiar laws of identity? Assume, for example, that $a = b$ and $b = a$ are "recarved" as two statements φ and ψ, respectively. Since the former two statements are logically equivalent, so too should be their "recarvings." This is ultimately a technical question, which accordingly receives a technical answer. So long as the relation on which we abstract is an equivalence, Frege demonstrates, there will be no problems as far as logic is concerned.[10] We shall return to a final and more philosophical objection in §2.7.

2.5 SOME FORMS OF REALISM

Let us digress briefly to reflect on some different forms of realism about mathematics. As explained in §1.4, object realism is the view that there exist mathematical objects. Another form of realism is concerned, not with the question of what there is, but with objectivity:

> **Truth-value realism.** Every well-formed mathematical statement has a unique and objective truth-value which is independent of whether it can be known by us or proved from our current mathematical theories.

Frege accepts both forms of realism. As we have seen, he compares mathematicians with geographers, who discover new continents whose existence and characteristics are independent of us.

---

[10] A relation is said to be an *equivalence* just in case it is reflexive, symmetric, and transitive.

Is there any relation between object realism and truth-value realism? In sciences other than mathematics, there certainly is. Statements about birds and planets, for example, are true or false because of the objects with which they are concerned and these objects' perfectly objective properties. Birds and planets thus figure in the explanation of the objectivity of statements concerned with such objects. The objects, in short, underpin the objectivity of the relevant discourse. What about mathematics? Platonism, we recall, holds that abstract mathematical objects are just as real as ordinary physical objects and play analogous roles in their respective sciences (cf. §1.4). This suggests the same order of explanation as above, namely that mathematical objectivity is underpinned by the existence of mathematical objects, which secure and explain the objective truth-values of statements concerned with such objects.

Frege rejects this order of explanation. He takes questions about the meaning of complete sentences to be explanatorily prior to questions about the reference of singular terms. On this view, the existence of mathematical objects is to be explained in terms of the objective truth-conditions of statements concerned with such objects rather than the other way round. Mathematical objects are never "given to us" directly, only via meaningful statements about them. A particularly radical form of this alternative order of explanation is encapsulated in Georg Kreisel's famous dictum that "the problem is not the existence of mathematical objects but the objectivity of mathematical statements."[11]

In sum, while Frege endorses both object realism and truth-value realism about mathematics, he stops short of full-fledged platonism. He reverses the platonist's view of the relative explanatory priority of mathematical objects and mathematical objectivity.[12]

---

[11] As reported by Dummett (1978b, p. xxxviii; see also 1981, p. 508). The remark of Kreisel's to which Dummett is alluding appears to be from a 1958 review (Kreisel, 1958, p. 138, fn. 1, which is rather less memorable than Dummett's paraphrase).

[12] Two other respects in which object realism can be weaker than platonism are discussed in §§4.4 and 5.2.

2.6 FREGE'S THEOREM

Frege's logicist project of reducing arithmetic to pure logic seemed to be going well. With his discovery of modern quantificational logic, there was reason to think that Kant had underestimated the sphere of analytic truths (cf. §§2.1–2.2). To show that the truths of arithmetic are analytic, Frege needed first to provide an analysis of these truths, and next to show that, thus analyzed, they can be proved from purely logical principles. We have seen the key elements of the requisite analysis, namely Frege's analysis of number ascriptions and his sophisticated account of Hume's Principle (cf. §§2.3–2.4). We shall now complete the technical part of the story, before returning to some philosophical concerns.

What exactly is the arithmetical theory that Frege wished to reduce to pure logic? The target theory is what is now known as *second-order Dedekind-Peano arithmetic*. This is an arithmetical theory that is formulated in a second-order language with three (for now) nonlogical symbols: a constant '0' for the number zero and predicates 'Pxy' and '$\mathbb{N}x$,' stating that x immediately precedes y and that x is a natural number, respectively. The theory has the following axioms:[13]

(1) $\mathbb{N}0$

(2) $\mathbb{N}x \wedge Pxy \to \mathbb{N}y$

(3) $\mathbb{N}x \wedge Pxy \wedge Pxy' \to y = y'$

(4) $\mathbb{N}y \wedge Pxy \wedge Px'y \to x = x'$

(5) $\mathbb{N}x \to \exists y\, Pxy$

(6) $\forall F \big(F0 \wedge \forall x \forall y (\mathbb{N}x \wedge Fx \wedge Pxy \to Fy) \to$
$\forall x(\mathbb{N}x \to Fx)\big)$

To reduce this theory to pure logic, Frege's first task is to provide a logical analysis of the three symbols still left unanalyzed. He begins by observing that zero is the number that applies to concepts under which no objects fall. This suggests that '0' be defined as '$\#x(x \neq x)$,' that is, as the number of objects that are

---

[13] A variant of the theory is presented in §11.2.

distinct from themselves. Next, what about the relation of one number immediately preceding another? This is simple enough to explain to a young child. Suppose you have some objects F and remove one of them, say a. Then the number of things left immediately precedes the number of things with which you started. This means that 'Pxy' can be defined as

$$\exists F \exists a \, (Fa \wedge x = \#u(Fu \wedge u \neq a) \wedge y = \#u \, Fu)$$

Finally, there is the concept of being a natural number. Informally, we often say that x is a natural number just in case $x = 0$, or $x = 1$, or …. But clearly, the ellipsis '…' would not have satisfied Frege. So he shows us how to do better. Say that a concept F is *hereditary*—abbreviated as $Her(F)$—just in case it is inherited under the predecessor relation:

$$\forall x \forall y (Fx \wedge Pxy \rightarrow Fy)$$

Frege proposes that '$\mathbb{N}x$' be defined as

$$\forall F (F0 \wedge Her(F) \rightarrow Fx)$$

In words: x is a natural number just in case it has every hereditary property had by zero.

Does Frege's approach to the first task succeed? As we have seen, his definitions use only logical vocabulary and the number-of operator '#.' So the answer will depend on whether this operator can be regarded as purely logical.

The second task is to show that Frege's definitions transform all the arithmetical axioms into theorems of logic. Unfortunately, Frege is not very explicit about how he understands "logic." He appears to be attracted to the idea that abstraction principles such as Hume's qualify as logical. Assume this idea is right. Then Frege does what it takes to complete his logicist project: he outlines proofs of all the axioms of Dedekind-Peano arithmetic from Hume's Principle and the mentioned definitions (§§74–82). Since his proofs have been spelled out and confirmed, the result is known as *Frege's theorem*. It is an amazing result. For more than a century now, informal arithmetic has been given a Dedekind-Peano style axiomatization, where the natural numbers are characterized in terms of their position in the number sequence. Frege's theorem shows that an alternative and conceptually completely different axiomatization is possible, where the natural numbers are characterized in terms of the concepts whose numbers they are.[14]

We conclude that, relative to the assumption that Hume's Principle qualifies as logical, Frege's logicism is a success.

2.7 DISASTER

Once this assumption is probed, however, the sense of success begins to crumble. One problem surfaces in §66. As we have seen, Hume's Principle assigns a meaning to every identity statement of the form "$\#x\,Fx = \#x\,Gx$." But it does not assign a meaning to identities of the form "$\#x\,Fx = t$," where 't' is a nonarithmetical singular term such as 'Julius Caesar' or a free variable. The worry may seem pedantic. Isn't it *obvious* that Caesar is distinct from every natural number? Perhaps. But even if so, this isn't something that Hume's Principle tells us. The principle remains silent on this class of sentences and thus fails to assign meanings to all sentences involving number terms.

Frege's fix was to provide an explicit definition of the numbers. To do so, he followed the lead of contemporary mathematicians such as Dedekind, who had discovered how to handle abstraction by equivalence classes rather than abstraction principles. Consider the case of directions. Instead of accepting directions as a distinct kind of object, we may identify the direction of a line l_1 with an equivalence class, namely the class of all lines l_2 that are parallel to l_1. It is straightforward to show that the principle (Dir) can now be derived and is thus not needed as an axiom. Frege suggested we apply the same strategy to numbers. There is no need to regard Hume's Principle as an axiom governing numbers understood as a distinct kind of object. It suffices to define the numbers as equivalence classes.

---

[14] In the terminology of §4.2, the Dedekind-Peano axiomatization views the numbers as finite *ordinals*, while Frege's approach views them as finite *cardinals*.

We may for example define $\#x\,Fx$ as the class of concepts that are equinumerous with F:

$$\#x\,Fx = \{G \mid F \approx G\}$$

Hume's Principle can now be derived and need not figure as an axiom. As before, the axioms of Dedekind-Peano arithmetic follow logically by means of Frege's theorem.

For Frege's revised strategy to work, however, we need a purely logical way to handle equivalence classes. Frege sought to achieve this by means of an abstraction principle for classes, namely his "Basic Law V," which states that two concepts have the same extension just in case they are coextensive. Let us write $\{x \mid Fx\}$ for the extension of the concept F. The law is then formalized as follows:[15]

(BLV) $\{x \mid Fx\} = \{x \mid Gx\} \leftrightarrow \forall x(Fx \leftrightarrow Gx)$

Frege's logicism had thus developed into a project of reducing all of arithmetic to Basic Law V, which he regarded as a logical principle. Whatever its other merits, this reduction has an appealing generality. Given any abstraction principle, we can define its abstracts as the appropriate equivalence classes and derive the principle from Basic Law V, using this definition. Basic Law V can thus be regarded as "the mother of all abstractions." Given this single abstraction principle, all the others follow.

But in 1902 disaster struck. Just as the second volume of his magnum opus, *Grundgesetze*, was going to press, Frege received a letter from the great logician and philosopher Bertrand Russell (1872–1970), who reported that he had "encountered a difficulty" with Frege's theory.[16] The "difficulty" is now known as *Russell's paradox*. As Frege knew, a membership predicate "$x \in y$" can be defined as "$\exists F(y = \{u \mid Fu\} \wedge Fx)$." That is, x is a member of y when y is the extension of a concept F under which x falls. Consider now the Russell class r, defined as $\{x \mid x \notin x\}$. Is r a member of itself? The answer is affirmative just in case r satisfies

---

[15] Is this law too subject to a "Caesar problem"? Frege eventually concluded that it is and proposed a solution (2013, §10).

[16] Their correspondence is reprinted in van Heijenoort (1967).

the membership criterion, namely not to be a member of oneself. This yields the contradictory answer that *r* is a member of itself just in case it is not. In a hastily written appendix, Frege proposed a fix, which was later shown to be inconsistent as well, assuming there are at least two objects. Around 1906, Frege seems to have given up on the reduction of arithmetic to logic. After this disappointing turn of events, Fregean logicism lay dormant for the better part of a century. But there have recently been attempts to revive parts of it. The leading proposal, due to Wright (1983), is to revert to the idea of adopting Hume's Principle as an axiom with a privileged, if not exactly logical, status. The fatal appeal to Basic Law V is thus avoided. We shall return to this proposal in Chapter 9, where we shall also discuss further Frege's intriguing idea that abstraction principles effect a "recarving of content."

SELECTED FURTHER READING

Relevant texts by Frege include the preface to *Begriffsschrift* (1879, available in Beaney (1997)) and *Foundations of Arithmetic* (1953, esp. introduction, §§1–4, 55–91, and 106–9). Boolos (1998) is an accessible general introduction, while Heck (1999) is a good introduction to Frege's theorem. Further important secondary literature includes Dummett (1991, esp. chaps. 10, 11, and 14), as well as Wright (1983, esp. §§i, ii, iii, v, and viii).

CHAPTER THREE

Formalism and Deductivism

3.1 A PURELY SYNTACTIC ALTERNATIVE?

As we have seen, Frege revived aspects of the ancient platonistic
conception of mathematics. In stark contrast to Plato, however,
Frege coupled this conception with an attempt to reduce arithmetic to logic. If successful, this reduction would have made
platonism more defensible, at least as concerns arithmetic. But
Frege's project ended in disaster. It is time to investigate whether
we can resist the push toward platonism that moved Frege. It
would be nice to give an account of mathematics that avoids all
philosophical speculation and is based solely on its incontrovert-
ible aspects.

One such aspect of mathematics is its concern with proof.
Indisputably, mathematicians prove theorems from axioms.
And since proofs consist of strings of symbols, they can be
described and studied in a purely syntactic way, thus avoiding
all philosophical speculation. Surely, one may think, this would
be preferable to speculation about abstract objects laid out in
"Plato's heaven." As Frege's colleague in Jena, the mathematician Carl Johannes Thomae, put it, "[t]he formal point of view
elevates us above all metaphysical difficulties" (quoted in Frege
(2013, §89)). This sentiment motivates the approaches to mathematics that will occupy us in this chapter.

Some definitions will be important throughout our discussion. First, we need to distinguish between syntax and semantics.
Syntax is the study of linguistic expressions abstracted from
any meaning they may have. The expressions are understood
as mere strings of characters. *Semantics*, by contrast, has
meaning as its primary concern. It is the study of the truth,
reference, and linguistic meaning that can attach to linguistic
expressions.

Frege's notion of a *formal system* enables a strictly syntactic conception of a proof as a string of formulas, which starts with a formula found on a list of axioms and every step of which involves either another such formula or else a formula obtainable from earlier members of the list by means of some syntactic operation from a list of rules of inference (cf. §2.1). Thus understood, a proof can be mechanically verified by a computer that knows nothing about the meaning of the formulas in question, if any. Reasoning in a formal system is thus like playing a game with formulas—which, as far as the game is concerned, might well be meaningless—subject to rules given by formally characterized axioms and rules of inference.

Formalism is the view that mathematics has no need for semantic notions, or at least none that cannot be reduced to syntactic ones. We shall in this chapter consider two versions of the view, as well as a kindred but more subtle view known as *deductivism*, which emphasizes mathematics' concern with the deduction of theorems from axioms.

Frege's relation to formalism is a curious one. He regarded the view as badly mistaken. Yet any proper articulation of the view requires Frege's own notion of a formal system, which makes available the strictly formal conception of proof. In fact, there is no tension here. Frege was a proponent of the axiomatic method and formalization, which make it possible to assess the logical correctness of a proof without taking into account the meaning of its formulas. But it is one thing to *be able to abstract from meaning* for certain purposes, and quite another to deny that the formulas *are meaningful at all*. Frege valued the former but abhorred the latter. We must distinguish between an expression's being meaningful, or "contentful" (from the German *inhaltlich*), and its being merely formal, or "empty." As Frege understood very clearly, meaningfulness is no obstacle to formalization.

Three final distinctions are needed. When discussing formalism, it is important to distinguish between use and mention. For example, ABBA is a pop group, while 'ABBA' is a name. In the former clause, the expression 'ABBA' is *used* to refer to a nonlinguistic item, while in the latter, the expression is *mentioned* to make a claim about this very expression. The use-mention distinction is particularly important as it applies to arithmetic. For example, '1' is a *numeral* (or number sign), whereas 1 is a *number*. Last, how many letters does the expression 'ABBA' contain? Depending on how we count, the answer may be either two or four. The expression contains two letter *types*, each of which has two occurrences. But it contains four letter *tokens*, that is, four printed characters, each with its distinct spatial location.

3.2 GAME FORMALISM

One version of formalism latches on to the comparison of a formal proof with a game played with syntactic expressions. According to *game formalism*, this is all there is to mathematics. That is, mathematics revolves around formal systems, which are syntactical games played with meaningless expressions. As Thomae put it:

> [A]rithmetic is a game with signs which one may well call empty, thereby conveying that (in the calculating game) they do not have any content except that which is attributed to them with respect to their behaviour under certain combinatorial rules (game rules). A chess player makes use of his pieces in a similar fashion. (Quoted in Frege (2013, §88))

If acceptable, this conception of arithmetic would indeed "elevate us above" most "metaphysical difficulties." As Frege observes, "[i]n formal arithmetic, we do not need to justify the rules of the game; we simply lay them down" (ibid., §89). There can be a question of justification only where a claim has been expressed, not where we are dealing with meaningless expressions.

While its potential advantages are clear, game formalism faces some serious challenges. The most immediate concern is that the analogy on which the view is founded is problematic. The language of mathematics appears to be meaningful. When you read an arithmetical sentence, for example, you seem to understand what it says. And this understanding seems to guide you when you later attempt to prove the sentence. Nowhere does the apparent meaningfulness of mathematical language come across more clearly than in connection with what we may call "preformalized" mathematics. Arithmetic, for example, was practiced for millennia before "the rules of the game" were finally made explicit in the 1880s in the form of the Dedekind-Peano axioms. The existence of highly developed mathematical reasoning in the absence of any clear set of "rules" suggests that mathematicians are often guided by the meaning of their sentences, not just by formal rules. Indeed, it is often by reflecting on what they *have in mind* that mathematicians are able to formulate and eventually adopt axioms that describe the relevant structure. For example, it is because the language of arithmetic is *about* what we shall later call a "simply infinite system" that the Dedekind-Peano axioms make sense and were eventually adopted (cf. §11.2).

Two further objections to game formalism were developed by Frege. The most famous one concerns how mathematics is "distinguished from a mere game" (2013, §90). Students are required to learn mathematics but not chess. And many grown-ups are paid good money for doing mathematical research, while few are paid to play chess. Why? Thomae's answer is that mathematics "can perform an essential service for us in the knowledge of nature" (quoted in §88). He is surely right that mathematical theories have useful applications. However, is this something that a game formalist is able to explain? Frege thinks not:

> If we do not look beyond [formal arithmetic's] boundaries, then its rules seem as arbitrary to us as those of chess. The applicability cannot, however, be a coincidence; but in formal arithmetic we spare ourselves any account of why we lay down the rules in exactly this way and not in any other. (ibid., §89)

As Frege observes, the applicability of arithmetic is not a co-incidence or something that we stumble upon once the game is up and running. It is an entirely predictable consequence of what arithmetical sentences mean. They express thoughts about

41

cardinality, which is a property that applies to the world via the concepts that specify what we are counting. As Frege puts it:

> Why can one make applications of arithmetical equations? Solely because they express thoughts. ... Now, it is applicability alone which elevates arithmetic above a game to the rank of a science. (ibid., §91)

A sophisticated game formalist might respond to Frege's objection as follows. "Okay, I concede that we have to account for the applicability of arithmetic. But why should this require that arithmetical theorems be true? It suffices that the rules governing the formal game of arithmetic are accompanied by bridge principles that relate arithmetical formulas to meaningful statements about the real world, and that these bridge principles are set up so as to ensure that an appeal to an arithmetical theorem will never lead us astray concerning the real world."

A simple example shows how the idea might be developed.[1] Assume we have the following:

(1) There is precisely one dog. (That is, $\exists x(\text{DOG}(x) \land \forall y(\text{DOG}(y) \to x = y))$.)
(2) There is precisely one cat. (That is, $\exists x(\text{CAT}(x) \land \forall y(\text{CAT}(y) \to x = y))$.)
(3) Nothing is both a dog and a cat. (That is, $\neg\exists x(\text{DOG}(x) \land \text{CAT}(x))$.)

Now we wonder how many things there are that are either a dog or a cat. A brilliant way to answer this question is by setting up *the applied arithmetic game*. First, we introduce bridge principles that link these zoological assumptions with some purely formal sentences of applied arithmetic:

(1^*) $\#x\,\text{DOG}(x) = 1$
(2^*) $\#x\,\text{CAT}(x) = 1$
(3^*) $\#x(\text{DOG}(x) \land \text{CAT}(x)) = 0$

Next, we invoke a trivial theorem of formal arithmetic:

(4^*) $1 + 1 = 2$

[1] The example is due to Putnam (1967b).

Finally, we adopt an inference rule that allows us to conclude from premises $\#x\, Fx = m$, $\#x\, Gx = n$, and $\#x(Fx \wedge Gx) = 0$ that $\#x(Fx \vee Gx) = m + n$. Using this rule, (1^*) through (4^*) entail

(C^*) $\#x(\text{DOG}(x) \vee \text{CAT}(x)) = 2$

We can now use a bridge principle to extract the real content of our formal conclusion (C^*):

(C) There are precisely two things that are either a dog or a cat. (That is, $\exists x \exists y((\text{DOG}(x) \vee \text{CAT}(x)) \wedge (\text{DOG}(y) \vee \text{CAT}(y)) \wedge x \neq y \wedge \forall z(\text{DOG}(z) \vee \text{CAT}(z) \rightarrow z = x \vee z = y))$.)

The sophisticated formalist response can now be expressed as follows. The arithmetical theorem (4^*) need not be true to make a useful contribution. All that matters is that (4^*)—and the rest of the applied arithmetic game—never lead us astray. And this guarantee can be had. As is well known, (C) follows in first-order logic from premises (1), (2), and (3). So the detour via the bridge principles and (4^*) can be eliminated. But this first-order derivation requires *a lot* of work. And things get *much* worse when the numbers involved are larger. The alternative derivation of (C)—over the bridge to formal arithmetic, an easy application of a trivial theorem, and back over the bridge to a claim about the real world—is far easier. In particular, this alternative strategy remains easy even when the numbers involved are larger.

How Frege might defend himself against this response is unclear. He might insist that the bridge principles endow the arithmetical sentences with some form of meaning, and that the response therefore goes beyond game formalism. But this does not go to the heart of the matter. A lethal opponent is no less dangerous if found to fight under a different flag. Some more robust defenses will be discussed in Chapter 7, where we shall examine a generalization (due to Hartry Field) of the response to Frege.

Frege's second objection to game formalism concerns the consistency of the games that mathematicians allegedly play. Surely, it is part of mathematicians' responsibility to ensure that the games they offer to lay people and scientists at the very least are consistent. Otherwise the games would be useless, and "the consumers" of mathematics would have a legitimate complaint against "the producers." Now, how can mathematicians discharge this responsibility? As Frege observes, just as we must distinguish between the game of chess and a mathematical theory of all the ways in which this game might be played, we must distinguish between a formal game and a mathematical theory of this game. This mathematical theory might for instance investigate whether we can derive '$0 = 1$' in the game, or whether the game can be extended in certain ways without collapsing to inconsistency. Frege makes the astute observation that we are here looking at contentful assertions about the game! That is, unlike the formal game itself, the mathematical theory of the game yields contentful theorems about what can and cannot be done in the game (2013, §93). This leaves game formalism with a dilemma. Either mathematicians can shirk their responsibility and refuse to vouch for the consistency of the theories that they promulgate, or else they can assume this responsibility but must then admit at least as much contentful mathematics as is required to develop the mathematical theory of the relevant games.

This is a powerful objection. The most sophisticated descendant of formalism that we shall encounter, namely that of Hilbert's program, opts for the second horn of the dilemma (cf. §4.1). In doing so, pure game formalism has been left behind.

3.3 TERM FORMALISM

As we defined it, formalism seeks either to banish all semantic notions from mathematics or else to reduce any such notions to purely syntactic ones. While game formalism pursues the former alternative, *term formalism* pursues the latter. Mathematical singular terms are now allowed to denote themselves.[2]

---

[2] This obliterates the use-mention distinction for mathematical singular terms because we have $0 = $ '0' and likewise for the other numerals.

The mathematician H. E. Heine (1821–81) expresses the view as follows.

> I take a purely formal point of view *by calling certain tangible signs numbers*, so that the existence of these numbers is not in question.
> (Quoted in Frege (2013, §87))

Historically, a major motivation for this view was to defend the existence of the imaginary number i, whose square is -1. If this number can be identified with a letter, its existence is surely not in question! But as Frege observes, this has the surprising consequence that numerals are "no longer external auxiliaries, like blackboard and chalk; rather, they comprise the essential components of the theory itself" (ibid).

What about the function symbols of arithmetic, such as '+' and '×'? These are defined by means of *rewrite rules* that allow us to rewrite numerals in a different way. Write 'S' for the syntactic operation that rewrites one numeral as the numeral that directly succeeds it. Thus, '$S(2)$' can be rewritten as '3.' We formalize this as $S(2) \rightsquigarrow 3$. Next, '+' and '×' are governed by the following rewrite rules:

$$m + 0 \rightsquigarrow m \qquad\qquad m \times 0 \rightsquigarrow 0$$
$$m + S(n) \rightsquigarrow S(m + n) \qquad m \times S(n) \rightsquigarrow (m \times n) + m$$

For example, '$2 + S(1)$' can be rewritten as '$S(2 + 1)$.' The rewrite rules enable us to compute with the numerals. We can for example demonstrate that $2 + 2 = 4$:

$$2 + 2 \rightsquigarrow 2 + S(1) \rightsquigarrow S(2 + 1) \rightsquigarrow S(2 + S(0))$$
$$\rightsquigarrow S(S(2 + 0)) \rightsquigarrow S(S(2)) \rightsquigarrow S(3) \rightsquigarrow 4$$

Term formalism is better placed than its more playful cousin to answer the problems discussed in the previous section. First, there was the problem that arithmetical formulas seem meaningful. As we have seen, term formalism allows such formulas to have a kind of content. For instance, '$2 + 2 = 4$' says that the former numeral can be reduced to the latter. This insight can be generalized. An arithmetical equation $s = t$ can be seen as

45

stating that there is a computation that reduces one of the terms to the other. Thus interpreted, the truth of the equation consists in there being such a computation.

The next problem concerns applicability. Let us reflect on how counting works. When we count, we produce a one-to-one correlation between the things to be counted and the numerals, starting with '1.' The numeral correlated with the last thing to be counted is said to "give the number" of the things in question. Notice that there is nothing in this account that a term formalist cannot accept. Next, we observe that each of our three arithmetical operators corresponds in a natural way to an operation on the things being counted. The successor operator 'S' corresponds to the operation of adding one thing to a collection. The reason is simple. By outputting the next numeral, this operation "gives the number" of the new and increased collection, on the assumption that the input numeral "gave the number" of the original collection. The addition sign '+' corresponds to combining two disjoint collections. To see this, assume that two numerals $\bar{m}$ and $\bar{n}$ "give the number" of two disjoint collections, respectively.[3] The rewrite rules governing '+' have been designed so as to ensure that '$\bar{m} + \bar{n}$' will "give the number" of the new and combined collection. Finally, the multiplication sign '$\times$' corresponds to a kind of repeated combination of disjoint collections. Consider some collections such that $\bar{m}$ "gives the number" of members of each of the collections, and $\bar{n}$ "gives the number" of the collections themselves. (It is useful to imagine the things as laid out in n rows, each with m members.) The rewrite rules governing '$\times$' have been designed so as to ensure that '$\bar{m} \times \bar{n}$' will "give the number" of the new collection that results from combining all of the given collections. In short, by granting to arithmetical terms and equations this simple form of meaning, as concerned with syntactical signs and operations, the term formalist makes strides toward an explanation of the applicability of arithmetic.

The term formalist is not yet out of the woods, however. Several problems remain. One is that the proposed analysis of the language of arithmetic isn't obviously correct. Consider for

[3] As usual, $\bar{m}$ is the mth numeral in the relevant notation system.

instance the identity sign '=,' whose meaning is ordinarily such that an identity statement is true just in case the singular terms that flank the identity sign refer to one and the same object. According to a term formalist, however, '$2 + 2 = 4$' is true although the two singular terms involved refer to distinct objects, namely themselves. The term formalist is thus committed to the view that the identity sign is ambiguous. It means something different in the context of arithmetic than when talking about the physical world.

Next, no account has been provided of the quantifiers that figure in arithmetical language. It might be tempting to rewrite a universal generalization $\forall n\, \varphi(n)$ as the infinite conjunction of instances $\varphi(0)$, $\varphi(1)$, and so on. But this is not an option, given our finite capacities. And no other rewrite rule is available that is guaranteed to reduce each quantified formula to some simpler formula.[4] This places severe limits on the potential scope of the term formalist account.

Finally, we need infinitely many numerals for the account to work. Should these numerals be understood as types or tokens? There is no evidence that infinitely many numeral *tokens* exist. As Frege observes, "[w]e have neither an infinite blackboard, nor infinitely much chalk at our disposal" (2013, §123). The numerals must accordingly be understood as *types*.[5] But linguistic types are abstract objects. For instance, although the type 'ABBA' is *instantiated* on the paper or screen that you are currently perceiving, it is not *located* there, or anywhere else for that matter. This abstractness means that some of the philosophical puzzles that we set out to avoid have not been completely eliminated. There may nevertheless be progress. Linguistic types are *quasi-concrete*, in the sense that they have canonical instantiations in space and time, in contrast to *pure abstract* objects, which do not.[6] And we may well get some epistemic handle on quasi-concrete objects via their canonical instantiations.

---

[4] However, we can get part of the way, as we shall see in §5.4.

[5] If not, a further problem arises as well, namely that '0 is larger than 1' threatens to come out true (not false, as it should) on the grounds that the former numeral token is larger than the latter.

[6] See Parsons (1980), to whom the distinction is due, as well as §8.4 below.

Let us take stock. By granting arithmetical expressions some form of meaning, term formalism makes some progress. In particular, it becomes possible to formulate a promising account of the applicability of at least quantifier-free arithmetic. But the account has limited scope and leaves some hard semantic and metaphysical questions unanswered.

3.4 DEDUCTIVISM AND THE STRUCTURAL APPROACH TO MATHEMATICS

Deductivism (sometimes also known as *if-then-ism*) is the view that pure mathematics is the investigation of deductive consequences of arbitrarily chosen sets of axioms in some formal and uninterpreted language. Just like game formalism, deductivism grants mathematicians a huge amount of freedom in the choice of which formal systems to investigate. The liberalism is most extreme in the case of game formalism, which holds that our choice is constrained only by a requirement of consistency— and perhaps by half an eye on which games the consumers of mathematics find useful when pursuing their various affairs. Deductivism is less extreme. It agrees that we are completely free to choose our axioms, subject to the minimal requirement of consistency. But deductivists deny that we have the same freedom concerning the rules that allow us to move from axioms to theorems. Deductivists insist that these rules be *deductively valid*. This insistence makes deductivism a far more powerful view than game formalism. As we shall see, it allows deductivists to assign some form of meaning to mathematical formulas and thus to explain their applicability. In order to explain these advantages of deductivism over game formalism, we first need some background from the history and methodology of mathematics.

Let us begin with the distinction between an abstract mathematical structure and its many realizations. Consider three playing cards, A, B, and C, and four permissible operations on these:

- e = do nothing
- α = flip B and C

- β = flip A and C
- γ = flip A and B

Let $x * y$ denote the result of first carrying out operation y, followed by operation x. We get the following "multiplication table," which describes what mathematicians call a *group*:

$*$	e	α	β	γ
e	e	α	β	γ
α	α	e	γ	β
β	β	γ	e	α
γ	γ	β	α	e

In fact, this abstract multiplication table arises in other contexts as well. Consider for instance the following operations on a globe, which can be seen to observe exactly the same multiplication table:

- e = do nothing
- α = rotate 180° about the north-south axis
- β = rotate 180° about same chosen axis in the equatorial plane
- γ = rotate 180° about the equatorial axis orthogonal to the previous

This pair of examples shows how two completely different systems of objects and relations can instantiate one and the same *abstract mathematical structure*. This abstract structure has been shown to have two different *realizations*.

Some terminological remarks are in order. Let a *system* be a collection of objects—known as *the domain*—on which are defined certain operations and relations. We leave open whether the collection, operations, and relations are understood set-theoretically or in terms of second-order logic (cf. §2.1). Next, we define what it is for two systems to be *isomorphic* (literally: "of the same structure"). The intuitive idea is that each system is a "mirror image" of the other. More precisely, two systems S and S' are isomorphic just in case there is a one-to-one mapping φ from the domain of S to that of S' which preserves all the operations and relations, in the following sense. Suppose that R and R' are matching dyadic relations of S and S', respectively. Then for any x and y in the domain of S, we have

$$R(x, y) \leftrightarrow R'(\varphi(x), \varphi(y))$$

And *mutatis mutandis* for other relations and for operations. For example, the two systems described above are easily seen to be isomorphic. Finally, two systems are said to *have (or realize, or instantiate) the same abstract structure* just in case they are isomorphic.

We have just talked about abstract structures as if these are entities of some sort. Although immensely natural, this talk need not be taken literally. To do so would be to prejudge a controversial question that will occupy us in Chapter 11. For present purposes, to say that two systems instantiate a common abstract structure may be understood as merely a shorthand for saying that the systems are isomorphic. Next, although the realizations in our two examples are physical, that is clearly not a requirement. The same abstract structure is also realized in various systems composed of abstract mathematical objects (assuming that such exist at all). Mathematicians often talk about a shared structure as "abstract" and its realizations as "concrete." We shall avoid this use of the terms, as it conflicts with our own definitions, which represent standard philosophical usage (cf. §1.4). Many realizations are abstract in our sense of this word, namely, they are nonspatiotemporal and causally inefficacious. We shall instead talk about the realizations as *particular*, and contrast this with the *general* structure that they instantiate.

Around the turn of the twentieth century, examples such as the one just described led to a new methodology in mathematics, pioneered in large part by the great mathematician David Hilbert (1862–1943). Consider for example geometry. Traditionally, Euclidean geometry was understood as an axiomatic theory of *physical space*, that is, the space in which we live our lives. With the emergence of non-Euclidean geometries in the 1830s, however, it was no longer obvious that physical space is Euclidean. If it is not, what would this mean for Euclidean geometry as a branch of pure mathematics? *Nothing whatsoever,*

Hilbert answered. Most other mathematicians agreed. Euclidean geometry describes a *mathematical space*, which may or may not be realized by our physical universe, but whose value and legitimacy as a piece of pure mathematics is independent of any such realization. The mathematical study of geometry is thus disentangled from the physical (and thus empirical) question about the structure of the space that we inhabit.

To facilitate this division of labor, Hilbert (1899) formulated an axiomatic theory of Euclidean geometry, which can be investigated independently of its traditional interpretation as concerned with physical space. The theory still uses the predicates 'point,' 'line,' and 'plane,' but only for heuristic purposes. There is no longer any requirement that these predicates mean what they used to mean when employed in theorizing about physical space. As Hilbert liked to emphasize, in a proper axiomatization of geometry, "one must always be able to say, instead of 'points, straight lines, and planes,' 'tables, chairs, and beer mugs' " (Hilbert, 1935, p. 403). Less flamboyantly put, Hilbert's point is that anything we assume in our proofs must be explicitly licensed by the relevant axiomatic theory. We may not go beyond what is so licensed, say by drawing on knowledge we possess about the extensions of the theory's atomic expressions on some preferred interpretation. The interpretation is to be left completely open.

This methodology can obviously be extended to other parts of mathematics as well. First we formulate axiomatic theories that describe, partially or completely, the abstract structure that we wish to study. Then we set out to prove theorems in these theories. Since the axioms hold for particular realization of the abstract structure in question, so will all of the theorems. Once again, this methodology enables a useful division of cognitive labor. Pure mathematics studies which theorems follow deductively from various axiomatic theories. And some appropriate empirical science studies which abstract structures are (approximately) realized in various natural systems that interest us.

3.5 DEDUCTIVISM ASSESSED

There can be no doubt that the methodology just described has become a legitimate and important part of mathematics. But the philosophical view known as *deductivism* is so impressed by the methodology that it takes the huge further step of claiming that mathematics should be based *entirely* on this methodology. As one of the view's foremost (though erstwhile) advocates put it, "the essential business of the pure mathematician may be viewed as deriving logical consequences from sets of axioms" (Putnam, 1975b, p. 41).

Let us begin by explaining how deductivists can ascribe meaning to the formulas of mathematics, which in turn enables an explanation of their applicability. The idea is that a formula can be understood as expressing that a certain structural property holds in any system of the sort described by the relevant set of axioms. Recall the simple group-theoretic example from the previous section. Here the formula $\forall x(x * x = e)$ can be understood as expressing (truly) that in any group-theoretic system whose structure is described by the given multiplication table, two repeated applications of any operation yields the identity operation. This is a generalization about the structural properties of a class of systems, some of which may be found in the physical world. Thus, by interpreting a mathematical formula in the mentioned way, we understand it as expressing an implicit universal generalization over systems with a certain abstract structure. Since this abstract structure may well be realized in reality, it is no surprise that the formula may lend itself to applications.

Does deductivism provide an acceptable philosophy of mathematics? Like many other philosophers, I believe the view takes a good idea too far. To explain some problems that the view faces, it is useful to distinguish between two different conceptions of axioms. On the traditional Euclidean conception, axioms are meaningful assertions, which serve as permissible starting points for deductive reasoning. By contrast, when a follower of Hilbert talks about the axioms of geometry or group theory, she means something completely different. Considered in isolation, the axioms of group theory are not assertions but comprise an implicit definition of some abstract structure. For instance, the associative axiom of group theory, $x * (y * z) = (x * y) * z$, simply lays down a condition that a system must satisfy in order to qualify as a group. Following the eminent logician and philosopher

Solomon Feferman (1928–2016), let us call the former type of axioms *foundational*, and the latter, *structural*.[7] As mentioned, no one would today deny the legitimacy of the abstract-structural methodology or its use of structural axioms. What the objections to deductivism aim to show is rather that mathematics cannot be based *entirely* on this methodology. In addition to structural axioms, mathematics requires some foundational axioms. We shall consider two arguments to this effect.

The first argument parallels one of our objections to game formalism. In addition to formal systems, we require some *contentful metamathematics*. We need a safe mathematical place to stand when reasoning about what does and does not follow deductively from some theory. In fact, deductive consequence can be approached in two different, but equivalent, ways.[8] On the semantic approach, a formula φ is said to be a consequence of a theory T just in case (roughly): for any system of objects and relations of which the axioms of T are true, φ is true as well. On the syntactic approach, φ is said to be a consequence of T just in case there is a proof of the former from the latter in some appropriate formal system. Either way, some contentful mathematics is needed to establish facts about deductive consequence and lack thereof. On the syntactic approach, for example, we require a countable infinity of quasi-concrete linguistic types and foundational axioms that suffice for reasoning about some simple syntactic operations on such types. These requirements are similar to those of arithmetic, which requires a countable infinity of natural numbers and axioms for reasoning about some simple arithmetical operations on numbers.

The second argument asks how we can find systems that realize the various abstract structures we find interesting. The question is important. For the standard way to show that some hypothesized abstract structure exists is by specifying a system that realizes the structure. By doing so, we show that the theory that characterizes (perhaps only partially) the hypothesized

---

[7] See Feferman (1999).
[8] The equivalence holds only for first-order theories (cf. §2.1). See Shapiro (2005a) for a useful introductory discussion of logical consequence.

structure is in good standing. But how can we find such realizing systems? Mathematicians typically proceed by "constructing" appropriate systems or models from some basic "building blocks," such as the natural numbers, by means of some operations, such as the formation of ordered pairs and equivalence classes. To show that an axiomatic theory of the real numbers is in good standing, for example, we "construct" a model that realizes the desired structure. This model typically consists of certain sets of rational numbers, which in turn are "constructed" from the natural numbers by means of further set-theoretic operations. But this and related "constructions" require foundational axioms that describe the "building blocks" and the operations that can be applied to these. Let us call this *the problem of model existence.*

Can the apparent need for foundational axioms be resisted? One option is to look outside of mathematics and seek realizations of the relevant abstract structures in the physical world. This is not what pure mathematicians do, however. And it is not hard to see why. If realizations of abstract structures had to be found in the physical world, mathematics would be hostage to empirical fortune. The reliance on the physical world would also conflict with the methodology that motivates deductivism, which seeks to disentangle pure mathematics from empirical questions and thus make the former independent of the latter.

Another option is to rely on structural axioms "all the way down." When the mentioned "constructions" appeal to natural numbers or sets, for example, perhaps we can regard this as just setting out certain deductive consequences of the relevant theory of numbers and sets, taking no stand on whether this theory in fact has models. But this too is unsatisfactory. The problem is nicely encapsulated by Putnam, who takes it to show that "mathematics is not *just* logic":

> It is a part, and an important part, of the *total* mathematical picture that certain sets of axioms are taken to describe *presumably possible* structures. It is only *such* sets of axioms that are used in *applied* mathematics. (Putnam, 1967b, p. 41)

54   Part of mathematicians' responsibility is to ensure that their theories are in good standing. If they promulgate theories that don't

even describe possible structures, the consumers of mathematics have a legitimate complaint. But the standard way to ensure that a theory is in good standing involves the "construction" of a model for it, which requires foundational axioms that describe the "building blocks" and permissible "means of construction."

Throughout the nineteenth century, the task of providing realizations of various abstract structures was increasingly assigned to set theory. This development was completed in the twentieth century. What makes this possible is the tremendous strength of modern set theory. The theory suffices to produce realizations of just about any abstract structure of mathematical interest. The resulting approach to mathematics, which is called *set-theoretic structuralism*, celebrates the structural methodology but supplements it with a strong theory of sets.[9] While this approach solves the problems of contentful metamathematics and model existence, it obviously gives rise to a new problem of explaining and justifying modern set theory.

SELECTED FURTHER READINGS

Frege's classic critique of formalism can be found in *Basic Laws of Arithmetic* Frege (2013, §§86–94, 113–14, 118–19, and 123–25). Weir (2015) provides a useful—and sympathetic—survey of formalism. The emergence of the structural approach to mathematics is usefully described in Burgess (2015). Putnam (1967b) provides an important (partial) defense of deductivism, while this view is criticized in Resnik (1980), pp. 54–75, 119–30.

---

[9] This is also the approach associated with the famous Bourbaki program of the mid twentieth century. See Bourbaki (1996).

Hilbert's Program

4.1 DIVIDE AND CONQUER

Let us pick up where the previous chapter left off. We found that deductivism, which draws much of its inspiration from the structural methodology pioneered by Hilbert, faces two problems. First, there is *the problem of contentful metamathematics*. In order to study the properties of formal systems or axiomatic theories, we need a contentful mathematical theory of syntactic strings and operations, and from a mathematical point of view, this theory is comparable with arithmetic.[1] Then, there is *the problem of model existence*. It is true that mathematics investigates the deductive consequences of axiomatic theories. But mathematics also needs its own foundational axioms in order to provide models for its various axiomatic theories, thus showing them to describe possible abstract structures.

The most sophisticated development of formalist ideas is that of Hilbert's program. Hilbert proposes to overcome the two mentioned problems by a brilliant strategy of divide and conquer. The way forward, he thinks, is to distinguish mathematics into two parts. *Finitary mathematics* is a contentful theory of finite and quasi-concrete syntactic types. Hilbert is particularly fond of numerals that take the form of strings of strokes; for example, '|||' is the third numeral. Such numerals are sequences of what we may call *Hilbert strokes*. Hilbert thinks that finitary mathematics and its foundational axioms can be accounted for using ideas from Kant and term formalism. *Infinitary mathematics*, on the other hand, is strong enough to describe all of the infinite structures that modern mathematics studies. This part of mathematics

---

[1] That is, each of the two theories can be "imitated" (or *interpreted*, to use the technical term) inside of the other.

can be regarded as a purely formal theory, Hilbert thinks, and when so regarded, can be accounted for by drawing on ideas from game formalism.[2]

We can now explain how this division of mathematics might help us overcome the problems that plague deductivism. As concerns the first problem, Hilbert concedes that we need a contentful metamathematics. But that is alright, he thinks, since the requisite metamathematical work can be done by his finitary mathematics, which has content. So far, so good. The problem of model existence seems harder, however. Finitary mathematics cannot possibly produce realizations of the vast, infinite structures that are studied in modern mathematics. Hilbert proposes a brilliant alternative. The semantic problem of model existence can be transformed into a purely syntactic problem about the *formal consistency* of the theories purporting to describe the relevant models. For example, instead of asking whether our theory of the real numbers has realizations, we ask whether the theory is formally consistent. And crucially, since the question of formal consistency turns on the existence of a finite and quasi-concrete derivation of $0 = 1$ or some other absurdity, this question falls within the range of Hilbert's finitary mathematics.

Of course, this brief outline of Hilbert's program raises many questions. We shall begin with the question of why Hilbert divides mathematics in the way he does. Given that he accepts some mathematics as contentful, why not extend this honor at least to some of the mathematics of the infinite? But that would be out of the question for Hilbert. While he is a staunch advocate of infinitary methods in mathematics, he denies that statements about the infinite can be understood as contentful. This ambivalence toward the infinite emerges clearly in the following passage:

---

[2] Arguably, Hilbert's skepticism about infinitary mathematics, regarded as a contentful theory, was largely of methodological nature: he sought to show that there is *no need* so to regard it. Our discussion will focus on Hilbert's program as described in " On the Infinite" (1926) and related writings. See Sieg (2013) for a discussion of nuances, such as the one just mentioned, and other Hilbertian programs.

> From time immemorial, the infinite has stirred men's *emotions* more than any other question. Hardly any other *idea* has stimulated the mind so fruitfully. Yet, no other *concept* needs *clarification* more than it does. (Hilbert, 1926, p. 185)

In the next two sections, we shall discuss some of the reasons Hilbert offers in support of his formal approach to the infinite.

4.2 SET THEORY AND THE PARADOXES

One reason for Hilbert and his contemporaries to be distrustful of the infinite was that the branch of mathematics that most explicitly had this as its concern, namely set theory, was threatened by paradoxes.

Modern set theory was developed in the late decades of the nineteenth century by the eccentric mathematical genius Georg Cantor (1845–1918). Most pre-Cantorian thinkers shared Aristotle's hostility to *the actual infinite*, that is, to the idea of infinite collections as somehow completed. Like Aristotle, the only infinity they accepted was *the potentially infinite*, where some process or operation can be repeated any number of times. For example, to consider the natural numbers as potentially infinite is to hold that necessarily, for any given natural number, it is possible to define or construct a larger natural number. It is a further step to regard the natural numbers as actually infinite, that is, as a completed collection every member of which already has a successor in this very collection.

One influential argument against the actual infinite was *Galileo's paradox*. Consider the function that maps a natural number to its square. This function associates every natural number with a unique square, and every square with a unique natural number. Assume for contradiction that the natural numbers can be regarded as a completed collection. Then our function defines a one-to-one correspondence between the entire collection of natural numbers $\{0, 1, 2, 3, \ldots\}$ and the subcollection consisting of all the squares, $\{0^2, 1^2, 2^2, 3^2, \ldots\}$. The latter collection is obviously smaller than the former, because it leaves out many numbers; in fact, it seems *much* smaller, for the further we proceed along the number line, the more thinly distributed the squares become. Our assumption thus implies that a collection can be one-to-one correlated with one of its proper subcollections and thus that a whole can have the same magnitude, or number, as one of its proper parts. But this contradicts an axiom recognized by Euclid, namely that a whole is always greater than its proper parts. We therefore conclude, by *reductio ad absurdum*, that the natural numbers cannot be regarded as a completed collection.

One of Cantor's boldest innovations was to take seriously the idea that a collection can have the same number as one of its proper subcollections. Perhaps the ancient Euclidean "axiom" is simply invalid for infinite collections.[3] Exploring this daring hypothesis, Cantor found that mathematics proceeds absolutely fine without this mistaken "axiom." He thus discovered a beautiful and rich mathematical theory of the infinite.

This theory retains the idea that collections that can be put in one-to-one correspondence have the same number—or *cardinal number*, as Cantor preferred to call it. In fact, this idea is now regarded as definitional of the notion of cardinal number. Cantor introduced another kind of number as well, namely what he called *ordinal numbers*. These have to do with *well orderings*, that is, linear orderings with the property that any subcollection of the ordering has a unique smallest element. Some examples will help to convey the idea. Consider first the integers, $\mathbb{Z} = \{\ldots, -2, -1, 0, 1, 2, \ldots\}$. These are not a well ordering; for instance, the negative numbers do not have a unique smallest element. Next, consider two copies of the natural number structure, laid out "head to tail":

$$0, 1, 2, \ldots, 0', 1', 2', \ldots$$

---

[3] In fact, the "axiom" fails for all and only the infinite collections. Say that a collection is *Dedekind infinite* just in case it can be put in one-to-one correspondence with one of its proper subcollections. Given a weak form of the axiom of choice, Dedekind infinity is equivalent to ordinary infinity, in the sense of being too large to be counted by a natural number.

This *is* a well ordering, as any subcollection must have a smallest element, either in the second copy of the natural number structure or else in the first. Cantor defines two well orderings to have the same ordinal number just in case they are isomorphic (that is, have the same abstract form). Thus, where the cardinal numbers specify the size or *cardinality* of collections, the ordinal numbers specify the order type of well orderings.

The smallest infinite ordinal, ω, is defined as the order type of the natural numbers. There are larger ordinal numbers as well. By appending a single object at the end of the sequence of the natural numbers, for example, we obtain a well ordering of order type $\omega + 1$; and the well ordering displayed in the previous paragraph has order type $\omega + \omega$. By iterating the operation of addition, Cantor showed how to define multiplication of ordinals, thus giving meaning for instance to $\omega \cdot \omega$. By iterating multiplication, he defined exponentiation, and thus for instance the number ω^ω. And this is only the beginning.

The cardinal number of the natural numbers was designated as $\aleph_0$ ("aleph-zero," after the first letter of the Hebrew alphabet). This is the smallest infinite cardinal. It might seem obvious that there are larger infinite cardinals as well. Consider $\aleph_0 + 1$, defined as the cardinal number of the collection of the natural numbers and one further thing. Or consider the cardinal number of the collection of the integers or of the rational numbers—surely each of these collections is more numerous than that of the natural numbers! But this line of thought is mistaken in the same way as Galileo's paradox. Although the natural numbers form a proper subcollection of each of the mentioned collections, Cantor showed that the former can be correlated one-to-one with each of the latter.[4] This means that all of these collections have the same cardinal number, namely $\aleph_0$.

---

[4] The hardest case concerns the rationals. Here is the key idea of the proof. Write the positive rationals in a two-dimensional table, with the numerators increasing step by step along one dimension and the denominators along the other. Then run through the table along the finite diagonals, beginning with the origin and "tagging" each new rational with a distinct natural.

So *are* there cardinal numbers larger than $\aleph_0$? Cantor gave an affirmative answer by examining *the powerset* of a given set A, written $\wp(A)$ and defined as $\{x \mid x \subseteq A\}$, that is, as the set of all subsets of A.

Cantor's Theorem. For any set A, its powerset $\wp(A)$ has more members than A itself.

The proof is as simple as it is ingenious. Assume the opposite. Then there is a function f from A onto its powerset.[5] Consider the diagonal set $\Delta = \{x \in A \mid x \notin f(x)\}$. Since f is onto, there must be some $\delta \in A$ such that $f(\delta) = \Delta$. Is δ a member of Δ? By the membership criterion for Δ, the answer is "yes" just in case $\delta \notin f(\delta)$. Since $f(\delta) = \Delta$, this yields

$$\delta \in \Delta \leftrightarrow \delta \notin \Delta$$

Since this is a contradiction, the theorem follows by *reductio ad absurdum*.

In particular, it follows that the cardinal number of $\wp(\mathbb{N})$ is larger than that of $\mathbb{N}$. Taking a step back, this is pretty shocking. Not only are there actually infinite sets, but infinite sets come in different sizes!

As a result of its remarkable strength, set theory provides realizations of practically all the mathematical structures of interest. An early but important example is the construction of the real numbers (as either Dedekind cuts or equivalence classes of Cauchy sequences of rationals). Based on this construction, the real numbers can be shown to have the same cardinality as the powerset of the naturals, that is, an infinity larger than that of the natural numbers. Cantor tried in vain to prove that this infinity is *the next one* after $\aleph_0$. This has become known as *Cantor's continuum hypothesis* and will be a central concern in our last chapter.

---

[5] A function g from A to B is said to be *onto* iff for each $b \in B$ there is $a \in A$ such that $g(a) = b$.

Sadly, set theory turned out to be prone to paradox. We have already encountered Russell's paradox, communicated to Frege in 1902 (cf. §2.7). The paradox concerns the set of non-self-membered sets, $r = \{x \mid x \notin x\}$. It is easy to prove the contradiction $r \in r \leftrightarrow r \notin r$. A host of other paradoxes were discovered around the same time. The paradoxes are fatal to so-called *naive set theory*, which is based on the following principles:

Naive Set Comprehension. Any formula $\varphi(x)$ defines a set $\{x \mid \varphi(x)\}$.

Extensionality. Coextensive sets are identical: $\forall z(z \in x \leftrightarrow z \in y) \rightarrow x = y$.

Since Frege and Dedekind were committed to naive set theory, they were badly affected. By contrast, Cantor denied, with at least some plausibility, that his approach suffered the same fate. Regardless of the exact list of casualties, the mathematical community clearly had a problem. Set theory, which increasingly had come to be regarded as a foundation for all of mathematics, had shown itself to be treacherous and fraught with difficulty. So it was hard not to agree with Hilbert's assessment that

> the present state of affairs where we run up against paradoxes is intolerable. Just think, the definitions and deductive methods which everyone learns, teaches, and uses in mathematics, the paragon of truth and certitude, lead to absurdities! If mathematical thinking is defective, where are we to find truth and certitude? (1926, p. 191)

In the 1920s, Hilbert therefore set himself the goal of giving set theory a secure foundation.

4.3 ARE THERE REALIZATIONS OF THE INFINITE?

One way to show a concept to be in good order is by showing that it is realized. Hilbert denies that this is an option with the concept of the infinite. He argues that we have no reason to believe this concept to be realized either within mathematics or outside.

Concerning the former, Hilbert draws inspiration from the elimination of *infinitesimals*, which were meant to be infinitely small quantities. Such quantities seemed to play an important role in early versions of the calculus. For example, what is the rate of change of the function $f(x) = x^2$ at $x = a$? The answer used to be to add an infinitesimal δ to the argument and investigate how this affects the slope of the function:

$$\frac{(a + \delta)^2 - a^2}{(a + \delta) - a} = \frac{(a^2 + 2a\delta + \delta^2) - a^2}{\delta} = 2a + \delta = 2a$$

As Berkeley famously observed, however, this answer is problematic. In the first two terms of our calculation, we are assuming that $\delta \neq 0$; otherwise, we would make the blunder of dividing by zero. But the transition from the third to the fourth term appears to assume that $\delta = 0$. This apparent doublespeak prompted Berkeley to mock that the analysts' δ is "the ghost of a departed quantity." The problem was conclusively resolved only in the work of Karl Weierstrass (1815–97), who let δ be an ordinary finite number distinct from 0 and considered the quantity $2a + \delta$ as δ approaches zero. This led Weierstrass to a precise mathematical definition of the notion of a limit, namely the epsilon-delta definition, which is used to this day.

As Hilbert observes, however, infinities remain elsewhere in mathematics:

> Nevertheless the infinite still appears in the infinite numerical series which defines the real number system and in the concept of the real number system which is thought of as a completed totality existing all at once. (1926, p. 183)

Don't these infinities show that the concept is realized? Hilbert thinks not, on the grounds that the relevant constructions are thoroughly set-theoretic in character and thus caught up in the crisis that affected set theory. As a last resort, we might try to find realizations of the infinite outside of mathematics. But we find no assurance there either, Hilbert claims. Certainly, no completed infinities are encountered in our experience. Nor do we have solid evidence that the physical world contains any completed infinities. It may be objected that physical space is infinite.

Hilbert admits that physical space is *unbounded*. However far we travel, we never reach a boundary that marks the end of space. But he observes that there are unbounded spaces which nonetheless are finite, such as a sphere, and that it is compatible with the general theory of relativity that our universe is such a space. Finally, appeal to the infinite divisibility of matter would be futile, since this is denied by quantum physics. Hilbert ends his fruitless search for realizations of the infinite with a dramatic conclusion:[6]

> Just as in the limit processes of the infinitesimal calculus, the infinite in the sense of the infinitely large and the infinitely small proved to be merely a figure of speech, so too we must realize that the infinite in the sense of an infinite totality, where we still find it used in deductive methods, is an illusion. (1926, p. 184)

4.4 HILBERT ON FINITISM AND POTENTIAL INFINITY

If the infinite is "an illusion," how should mathematics be developed? Let us begin with those parts of mathematics that have meaning or content. This content must obviously be concerned only with finite objects. What are these objects? And how do we come to know them?

As mentioned, Hilbert's answer draws on both Kant and earlier term formalism. He claims that "[t]he subject matter of mathematics is … the concrete symbols themselves whose structure is immediately clear and recognizable" (1926, p. 192). In particular, arithmetic is concerned with the Hilbert strokes and their arrangement, where these strokes serve as proxies for the natural numbers. This answers one of our questions. The contentful part of mathematics is concerned with quasi-concrete syntactical objects.

As concerns epistemology, Hilbert attempts to use Kant's account of our knowledge of the quasi-concrete objects that are

---

[6] Unlike Putnam (1967a) and Hellman (1989), Hilbert seems not to have considered whether the concept of the infinite might be shored up by *possibly* having realizations (cf. §11.2).

the subject matter of finitary mathematics. The basic idea is that quasi-concrete types can be perceived or intuited in a way that isn't obviously empirical and thus doesn't render the resulting mathematical knowledge *a posteriori*. On Kant's account, this form of "pure intuition" merely delivers facts about the form of our own sensibility (cf. §1.6). This account presupposes Kant's transcendental idealism, and it is unlikely that Hilbert was prepared to go that far. But there are alternative ways to develop Hilbert's basic idea. Some accounts of mathematical intuition that are broadly Kantian but not idealist will be discussed in §8.4.

What is the scope of Hilbert's finitary mathematics? The most widely accepted answer, due to William Tait, is that Hilbert equates finitary mathematics with *primitive recursive arithmetic*, first explicitly formulated by Thoralf Skolem (1887–1963).[7]

How a finitist such as Hilbert can accept even this elementary arithmetic theory as contentful may seem puzzling. After all, he claims that the infinite is "an illusion," and even this elementary arithmetical theory is committed to infinitely many natural numbers! The answer is that Hilbert understands the natural numbers—or the Hilbert strokes that represent them—as merely potentially infinite, not as an actually infinite collection, which *would* conflict with his finitism. The distinction between two kinds of infinity cannot be dismissed as empty rhetoric. For the focus on potential infinity leads Hilbert to disallow quantifiers that purport to range over all the natural numbers. Consider Euclid's theorem that, for every prime number p, there is another prime that is larger than p but smaller than or equal to $p! + 1$ (where $p!$ is defined as $1 \cdot 2 \cdot 3 \cdot \ldots \cdot p$). This statement is unproblematic from a finitist point of view. Since the existential generalization has an upper bound, it can be interpreted as a finite disjunction of its instances. Things are very different with *unbounded* generalizations over the natural numbers, such

---

[7] This theory uses no quantifiers but allows signs for all primitive recursive functions, such as addition, multiplication, and exponentiation. Its axioms are those of first-order Dedekind-Peano arithmetic (cf. §11.2), except that the induction schema is restricted to quantifier-free formulas, but supplemented with all the recursion equations for the mentioned functions.

65

as the statement that either $p + 1$ or $p + 2$ or $p + 3$ ... or (ad infinitum) ... has a certain property. This statement involves "an infinite logical product" (or disjunction), and

> [this] extension into the infinite is, unless further explanation and precautions are forthcoming, no more permissible than the extension from finite to infinite products in calculus. Such extensions, accordingly, usually lapse into meaninglessness. (1926, p. 194)

For an example of the kind of illegitimate extension Hilbert has in mind, consider $(-1)^\omega$. Observe that $(-1)^\omega = (-1)^{1+\omega} = (-1) \cdot (-1)^\omega$. Since $x = (-1) \cdot x$ only when $x = 0$, this implies that $(-1)^\omega = 0$. But this conclusion is unacceptable because a product equals 0 only when one of the factors does. So we are forced to conclude that the infinite product is undefined or "meaningless."

It might be objected that arithmetic would be crippled without the ability to quantify over all the natural numbers. Consider, for example, the commutative law of addition, which states that $a + b = b + a$ for any natural numbers a and b. Hilbert responds by articulating a finitistically acceptable notion of *schematic generality*, which allows us to simulate some of the unavailable quantifications. For example, the mentioned law can be understood as stating that for any two numerals that might be produced, the associated instance of the equation is true; and each of these instances is finitistically acceptable. More generally, any arithmetical formula with only free variables can be understood as a schematic generalization over any numerals that might be produced and serve as values of the variables.

Regrettably, the use of schematic generalities to simulate quantification is severely limited; for example, we cannot simulate negated universal quantifications. To see this, consider a schematic generality $\varphi(a)$, which simulates the universal generalization $\forall n\, \varphi(n)$. We would like to simulate its negation, $\neg\forall n\, \varphi(n)$. Our best shot is the negation of the schematic claim, namely $\neg\varphi(a)$. But this simulates a universally generalized negation, $\forall n\, \neg\varphi(n)$, not the desired negated universal generalization, $\neg\forall n\, \varphi(n)$. Hilbert therefore concludes that a schematic generality "is from our finitary perspective *incapable of negation*" (1926, p. 194). This has dramatic consequences:

> From our finitary viewpoint, therefore, we cannot argue that an equation like the one just given, where an arbitrary numerical symbol occurs, either holds for every symbol or is disproved by a counter example. Such an argument, being an application of the law of excluded middle, rests on the presupposition that the statement of universal validity of such an equation is capable of negation. ... In short, the logical laws which Aristotle taught and which men have used ever since they began to think do not hold. (Ibid.)

This objection to classical logic anticipates a cornerstone of the intuitionistic conception of mathematics, to be discussed in the next chapter.

Hilbert's analysis of the potential infinity of the natural numbers illustrates one way in which object realism (which holds that there exist mathematical objects) is weaker than platonism (which adds the loosely defined claim that these objects are "just as real as" physical objects). Suppose we established that each of some infinite range of physical objects exists. This would establish that the entire range exists and can serve as a domain for the quantifiers of classical logic. But Hilbert denies that the same holds for the natural numbers. Although *each* number exists, there is no *completed totality* of numbers over which classical quantification is defined. The reason is that the existence of a natural number is *merely potential*. The existence of a natural number is in this respect less robust than that of a physical object.

To make the argument more explicit, let us formalize the claim that a number exists not by '$\exists n$' but by '$\lozenge \exists n$.'[8] The claim that every number has a successor is then formalized as

(4.1) $\Box \forall m \lozenge \exists n \, \text{SUCCESSOR}(m, n)$

Given the existence of zero, this claim ensures that each number exists—in the potential sense expressed by '$\lozenge \exists$.' By contrast,

---

[8] As usual in philosophy, '$\Box$' and '$\lozenge$' represent necessity and possibility, respectively.

the existence (in said sense) of a completed totality of numbers would require

$$(4.2) \qquad \Diamond \forall m \exists n \; \text{SUCCESSOR}(m, n)$$

The key observation is now that (4.1) does not entail (4.2). Although we can always go on to produce a successor—by appending another stroke to some given sequence of Hilbert strokes—it does not follow that it is possible to complete this infinite process of appending further strokes.[9]

4.5 THE METHOD OF IDEAL ELEMENTS

The scope of contentful mathematics is severely limited if Hilbert is right. The right response, one might have thought, is to cut mathematics down to size. (This is Brouwer's response, as we shall see in the next chapter.) Nothing could be further from Hilbert's view. He insists that "[n]o one shall drive us out of the paradise which Cantor created for us" (1926, p. 191). Moreover, "[t]aking the principle of excluded middle from the mathematician would be the same, say, as proscribing the telescope to the astronomer or to the boxer the use of his fists" (Hilbert 1927, p. 476). Being a mathematical heavyweight himself, Hilbert speaks with authority. But how can we salvage "Cantor's paradise"? After all, Hilbert believes the paradise to lie beyond the scope of contentful mathematics. The proposed solution is that infinitary mathematics *need not be contentful in order to be justified.*

This orientation is often known as *working realism.* Hilbert undertakes to defend the methods typically associated with a platonistic outlook, such as reasoning about completed infinite collections and unrestricted use of classical logic. The defense must not rely on a platonistic philosophy but must be internal to mathematics, drawing only on its finitary and contentful parts.

---

[9] It is useful to represent possibilities by possible worlds. The nonentailment can then be proved by considering a system of possible worlds that contain arbitrarily large—but always finite—initial segments of the natural numbers.

This feat is to be achieved by *the method of ideal elements*, which we shall now explain.

Consider the introduction of complex numbers in the sixteenth century in order to obtain roots of equations that would not otherwise have any. For example, the imaginary unit i was introduced as a root of the equation $x^2 + 1 = 0$. These new numbers caused great controversy. Do such strange numbers really exist? The method of ideal elements is an attempt to sidestep this metaphysical question. The new numbers are a form of useful fiction. We expand our language to allow talk of such numbers and formulate a theory in this language that describes them. But this theory need not be true in order to be useful. It suffices that the theory is precisely specified and doesn't enable us to prove any new claims about the "old" numbers that we accept with full metaphysical earnestness.[10] Closer to Hilbert's own time, a variety of ideal elements came to be accepted in this way. In the early nineteenth century, for example, geometers postulated "points at infinity" as imagined points of intersection of parallel lines, thus giving rise to a beautiful theory of projective geometry.

Hilbert's aim is to treat the infinite in the same manner:

> [W]e conceive mathematics to be a stock of two kinds of formulas: first, those to which the meaningful communications of finitary statements correspond; and secondly, other formulas which signify nothing and which are the *ideal structures of our theory*. (1926, p. 196)

That is, we can formalize the language and theory of infinitary mathematics and treat it as an ideal superstructure, which can be studied without regard to its meaning, by means of our contentful finitary mathematics. As Hilbert reminds us,

> [T]here is just one condition ... connected with the method of ideal elements. That condition is a *proof of consistency*, for the extension of a domain by the addition of ideal elements is

---

[10] Hilbert even compares his ideal elements with Kant's "ideas of reason," (e.g., at Hilbert (1926, p. 201)).

legitimate only if the extension does not cause contradictions to appear in the old, narrower domain, or, in other words, only if the relations that obtain among the old structures when the ideal structures are deleted are always valid in the old domain. (Ibid., p. 199)

The potential payoff that Hilbert envisages is great. The mathematical methods typically associated with platonism might be justified without actually having to be a platonist. This justification would take the form of a consistency proof for the infinitary theories that employ such methods, but where this consistency proof is carried out in Hilbert's own contentful finitary mathematics.

With this strategy in place, all that remained was the purely mathematical task of producing a finitary proof of the consistency of ideal mathematics. Since Göttingen under Hilbert was home to some of the best mathematicians in the world, there was reason for optimism.

4.6 GÖDEL'S INCOMPLETENESS THEOREMS

This sense of optimism received an unexpected blow in September 1930, when the 24-year-old Kurt Gödel (1906–78) announced a theorem at a conference in Königsberg. His result is now known as *the first incompleteness theorem*.[11] Incorporating an improvement due to Rosser, the theorem says, roughly speaking, that any consistent formal system F in which a certain amount of elementary arithmetic can be carried out is incomplete. That is, there are statements of the language of F which can neither be proved nor disproved in F. The philosophical upshot is that no single formal system can prove all the truths of mathematics. This was surprising and alarming to mathematicians such as Hilbert, who had great faith in the

---

[11] Our discussion of the incompleteness theorems will be brief, as there are other good expositions. See, e.g., Boolos et al. (2007) or Raatikainen (2015).

power of formal systems to capture our informal mathematical reasoning.

Two months later, things took a turn for the worse. Gödel and John von Neumann independently discovered a corollary, now known as *the second incompleteness theorem*. Let F be as above. Then F cannot prove its own consistency. This result is disastrous for Hilbert's program. It means that carrying out the task that remained for Hilbert and his associates is mathematically impossible. Finitary mathematics cannot prove its own consistency, let alone that of a stronger system. As a result of this theorem, Hilbert's program is now almost universally regarded as dead.

In fairness to Hilbert, I hasten to add that his program has nevertheless borne many fruits, even if not the particular one that Hilbert most desired. His program launched proof theory as a branch of logic, which continues to provide many deep results about the relations between formal theories. Even if the second incompleteness theorem undermines Hilbert's attempt to use a weak theory to prove the consistency of a strong one, it is still possible to prove the consistency of one theory *assuming the consistency of another theory*. Such results can be illuminating, for instance when the two theories enjoy different kinds of evidence.[12]

The most important post-Gödelian attempt to revive formalism is due to Haskell Curry (1900–1982), who insists that "a proof of consistency is neither a necessary nor a sufficient condition for acceptability" (1954, p. 205). Curry thus advocates a kind of formalism without a safety net. The main problem with this view surfaced already in our discussion of game formalism and deductivism (cf. §§3.2 and 3.5). Mathematicians have a professional responsibility to exercise due care to ensure that their theories are consistent. The various consumers of mathematics would have a legitimate complaint if mathematicians were found to peddle damaged goods. True, the second incompleteness theorem limits mathematicians' ability to *prove* consistency without relying on assumptions

---

[12] See Feferman (1988) for an overview of these *relative consistency* results.

whose consistency can in turn be doubted. But the desired assurance need not take the form of a proof. Another option is to abandon strict formalism, recognize that at least some parts of mathematical language have content, and—drawing on this content—to argue that the relevant mathematical axioms are true and thus also consistent. This is what Hilbert does in his broadly Kantian account of finitary mathematics; a related attempt will be discussed in the next chapter. More ambitious attempts to draw on the content of mathematical language to provide evidence for the truth of its axioms will be considered in later chapters, especially 8, 10, and 12.

SELECTED FURTHER READINGS

The essential text on Hilbert's program is "On the Infinite" (1926), although "The Foundation of Mathematics" (1927) is also useful. von Neumann (1931) provides an accessible presentation and defense of Hilbert's program. Tait (1981) develops a highly influential analysis of finitism. Two valuable recent discussion of Hilbert's program by leading contemporary scholars are the survey by Zach (2015) and the monograph by Sieg (2013).

Intuitionism

5.1 LIVING WITHIN ONE'S MEANS

The intuitionist approach to mathematics, developed by the great Dutch mathematician L.E.J. Brouwer (1881–1966), provides an interesting contrast to Hilbert's program. Brouwer's view of contentful, finitary mathematics has much in common with Hilbert's—indeed, the former may well have influenced the latter.[1] As concerns finitary mathematics, Brouwer's main complaint is only that Hilbert does not develop this form of mathematics far enough. But a deep difference between the two thinkers emerges when we turn to infinitary mathematics. Brouwer not only rejects Hilbert's purely formal conception of these parts of mathematics but in fact goes to the opposite extreme by claiming that language is utterly irrelevant to mathematics, whose proper concerns are purely mental constructions.

This deep difference means that the two thinkers were affected by Gödel's second incompleteness theorem in very different ways. As we have seen, Hilbert sought a great philosophical bargain, namely infinitary mathematics—albeit formalistically construed—for the epistemic and metaphysical price of just finitary mathematics. Gödel's theorem therefore came as a horrible surprise to Hilbert, as it meant that no such bargain is available. By contrast, Brouwer seems to have been unsurprised by the incompleteness phenomenon and may well have relished the destruction it wrought on Hilbert's program.[2] The moral that Brouwer drew from the theorem was that we must learn to live within our means. More precisely, we must develop, as far as possible, the contentful, finitistic part of mathematics. This is

---

[1] Brouwer certainly thought so. See Brouwer (1928).
[2] See van Atten (2014, §3.5).

the fundamental aim of intuitionism. As Brouwer's student and fellow intuitionist Arend Heyting (1898–1980) put it, "[w]e …are interested not in the formal side of mathematics, but exactly in that type of reasoning which appears in metamathematics; we try to develop it to its farthest consequences" (1956, p. 68).

As we shall see, Brouwer's frugal approach meant a substantial departure from the classical mathematics that Hilbert helped shape and whose consistency he unsuccessfully tried to prove. It would be wrong, however, to think of intuitionism as all about depriving us of parts of the classical mathematics that we have come to appreciate. On the contrary, intuitionism is one of the approaches to the philosophy of mathematics that has done the most to build up new mathematics, not just to tear down old. This is perhaps unsurprising, given that Brouwer was a brilliant mathematician, who made important advances in classical mathematics before his attention shifted to the development of an intuitionistic alternative. Indeed, Brouwer is regarded as the founder of modern topology, to which he continued to contribute after his turn to intuitionism.

Because of its combination of frugality and creative innovation, intuitionism combines both reactionary and progressive elements. Let us begin with the former. Before the nineteenth century, much of mathematics was concerned with constructions. In geometry, for instance, mathematicians talked about *drawing* a line between any two points and *extending* any given line as far as one pleases. Following Hilbert's axiomatization of geometry, we now use the nonconstructive language of modern logic to say that for any two points, there *exists* a line connecting the points. Since this line is assumed to be infinitely long in each direction, there is no room for later "extensions."

Might this shift from the dynamic vocabulary of construction to the static vocabulary of existence be merely a terminological change? To think so would be mistaken. The shift had pervasive consequences for the methodology of mathematics. The dispute about nonconstructive existence proofs provides a good example. In order to prove an existential generalization, mathematicians traditionally needed to construct an instance, or at least provide an algorithm for doing so. This changed in the course of the nineteenth century. Consider the question of whether there are irrational numbers a and b such that a^b is rational. Now, $\sqrt{2}^{\sqrt{2}}$ is either rational or not. If the former, then $a = b = \sqrt{2}$ yields numbers with the desired property. If the latter, then $a = \sqrt{2}^{\sqrt{2}}$ and $b = \sqrt{2}$ yield such numbers, because we have

$$(\sqrt{2}^{\sqrt{2}})^{\sqrt{2}} = \sqrt{2}^{(\sqrt{2} \cdot \sqrt{2})} = \sqrt{2}^2 = 2$$

What troubled traditionalists is that the proof doesn't settle *which* pair of numbers has the desired property. To settle such questions, a proof must not rely on *the law of excluded middle—* that is, $\varphi \vee \neg\varphi$—which is used in the above argument to claim that $\sqrt{2}^{\sqrt{2}}$ is either rational or not.

Another controversial development concerned the use of infinitary methods. As we have seen, most mathematicians before Cantor rejected the idea of the actual infinite and accepted only the potentially infinite (cf. §4.2). As a result of Cantor's seminal work, however, it was widely recognized that actually infinite sets are extremely useful for proving results about more traditional mathematical objects. Consider the theorem that there are transcendental numbers.[3] This was painstakingly proved by Liouville in 1844. Using Cantor's theory of the infinite, a routine cardinality argument suffices to show that nearly all real numbers are transcendental.[4] Again, mathematical traditionalists were troubled, as Cantor's proof makes extensive use of his theory of infinite sets.

Might the intuitionists' criticism be dismissed as merely a yearning for the return of the *ancien regime*? To do so would be rash. The intuitionists buttressed their case by observing that set theory had been discredited by the paradoxes (cf. §4.2). After Gödel's theorems, they could add that the use of infinitary

---

[3] A real number x is said to be *transcendental* just in case there are no integers $a_0, a_1, \ldots, a_n$ such that $a_0 + a_1 \cdot x + \ldots + a_n \cdot x^n = 0$.

[4] In briefest outline: While there are uncountably many real numbers, there are only countably many equations with integer coefficients, and thus also only countably many solutions to such equations.

methods and application of classical logic to the infinite cannot be justified on finitistic grounds alone but rely on problematic metaphysical assumptions. In particular, these methods seem to presuppose a platonistic conception of an objective mathematical universe comprising vast infinities of abstract objects.

5.2 BROUWER ON MENTAL CONSTRUCTION

Suppose the intuitionists are right that classical mathematics relies on problematic metaphysical assumptions. To complete their case, they need to provide an alternative foundation for their own favored finitistic mathematics, which shows that this form of mathematics avoids the problematic assumptions. Brouwer's alternative foundation is based on the view that "there are no non-experienced truths" in mathematics (1949, p. 90). Echoing Berkeley's famous slogan that "to be is to be perceived," the intuitionists might be seen as claiming that in mathematics *to be is to be constructed*. This is a radical departure from platonism known as *antirealism*. While there are mathematical objects, these are our own mental constructions and thus not independently real in the same way as physical objects.

According to Brouwer, these mental constructions have nothing to do with language. Mathematics is an essentially subjective activity. Language is immaterial to mathematics and merely an optional extra for those mathematicians who wish to communicate their mental constructions to others.[5]

As an avowed antirealist about mathematics, it is no surprise that Brouwer traces his view back to Kant. But he cannot accept Kant's view unmodified. Like most of his contemporaries, Brouwer took Kant's view of geometry to be refuted by the discovery of non-Euclidean geometries. How can the *a priori* truths of Euclidean geometry be underwritten by our forms of sensibility if there are alternative truths about non-Euclidean geometries, which seem equally *a priori*? According to Brouwer, however, it would be wrong to dismiss Kant's philosophy of

[5] See, e.g., Brouwer (1913, p. 81).

mathematics entirely. For this philosophy "has recovered by abandoning Kant's apriority of space but adhering the more resolutely to the apriority of time" (1949, p. 80). More specifically, we must consider

> the falling apart of moments of life into qualitatively different parts, to be reunited only while remaining separated by time, as the fundamental phenomenon of the human intellect. ... This intuition of two-oneness, the basal intuition of mathematics, creates not only the numbers one and two, but also all finite ordinal numbers, inasmuch as one of the elements of the two-oneness may be thought of as a new two-oneness, which process may be repeated indefinitely. (ibid.)

So although Kant was wrong about geometry, he was right about arithmetic. In particular, we can derive from Kant the "basal intuition" of "two-oneness." Consider a moment of time and then wait a little. Now you have another moment of time, distinct from the original one. This "basal intuition" serves as Brouwer's mental analogue of Hilbert's operation of appending another Hilbert stroke. Each operation converts a representation of a natural number into a representation of the next number. For instance, if you have just observed ten successive fallings apart of a moment of your life into two, then by waiting a little and observing the falling apart of the last of these moments as well, you obtain a representation of the number twelve.

What are we to make of all this? Brouwer's subjectivist antirealism has left many readers exasperated. As even a sympathetic commentator remarks, Brouwer's "homespun phenomenology and ontogenesis may well grate upon some ears" (Posy, 2005, p. 331). Let us calmly try to identify and assess the problems. The most pressing question is how to understand the notion of a mental construction. Brouwer is sometimes taken to be concerned with constructions in *each individual person's mind*. This interpretation results in a so-called *psychologistic* view of arithmetic, which identifies the numbers with representations in each individual's mind. And this, in turn, makes Brouwer vulnerable to one of Frege's classic objections. On the psychologistic view, Frege observes, your natural numbers are different from mine; after all, the former exist only in your mind, and the latter, only in mine. To explain the fact that mathematicians communicate successfully and to give mathematics some degree of objectivity—or at least intersubjective validity—the account has to be supplemented with an explanation of how you and I manage to coordinate our mental constructions. As we have seen, however, Brouwer is uninterested in how mental constructions are communicated.

Other commentators deny that Brouwer should be understood as concerned with empirical goings-on in individual people's minds. Indeed, Brouwer's invocation of Kant makes it reasonably clear that whatever he takes himself to be doing, it is not empirical psychology. A similar lesson emerges from Heyting, who qualifies the mind-dependence of mathematics to avoid the radical subjectivism that Frege deplores: "[e]ven if [the integers] should be independent of individual acts of thought, mathematical objects are by their very nature dependent on human thought" (1931, p. 53). That is, the natural numbers do not depend on your mind or on mine, although they do depend on human thought in general. However, this is merely a gesture toward a desired conclusion, not an argument that takes us there. Can we do better?

One option is to seek inspiration from the phenomenonology of Edmund Husserl (1859–1938). Perhaps Brouwer is concerned with a "transcendental subject" in something like Husserl's sense, where this involves "aspects of subjectivity that are the same for everyone precisely in virtue of being a subject" (van Atten, 2004, p. 80). An interpretation along these lines would, if defensible, enable intuitionists to respond to some standard objections. Consider, for example, a reclusive genius who proves a theorem but dies before conveying it to anyone else. Are intuitionists committed to the implausible claim that the theorem ceases to be true with the death of our genius? Perhaps not, since on the present interpretation they take mathematical truths to be independent of "individual acts of thought."

Some related problems would still remain, however. By making mathematics dependent on human thought, even if not on individual acts thereof, we would still ascribe incorrect temporal and modal properties to mathematical truths. As Frege famously argued, all mathematical truths now known were true already before we discovered them, indeed even before the existence of the human species (cf. §1.3). This is why it is permissible to apply mathematics, as we now have it, to theorize about not only the present and future, but also the distant past. It is hard to see how claims that were true even before the arrival of the human species can be said to be "dependent on human thought." An analogous point applies to modality rather than time. The truths of mathematics would have been true no matter what had happened, indeed even if the human species had never existed. This is why it is permissible to apply mathematics, as we actually have it, to theorize not only about what is actual but also about all counterfactual scenarios. It is hard to see how such truth can be said to depend on "human thought."

To address these problems, intuitionists appear to have no choice but to loosen the link between mathematical truth and actual human possession of a proof. Perhaps it suffices for truth that the claim in question is *humanly provable* rather than actually proved? The critical question is then what is meant by "humanly provable." How are we to extrapolate from what we have actually proved to what we are able to prove? There is a danger that such an extrapolation will fall back on a realist conception on which the relevant facts were "there all along," waiting to be discovered.

5.3 Intuitionistic Logic

As mentioned, intuitionists reject some of the laws of classical logic. The first systematic investigation of the alternative intuitionistic logic was undertaken by Heyting. Of particular importance is the so-called *BHK interpretation* of the connectives and quantifiers, developed by Heyting and Kolmogorov, and inspired by Brouwer. This interpretation has proof as its central notion, rather than truth, as in the standard semantics for classical logic. The interpretation is based on the following proof conditions:

- a proof of $\neg A$ is a derivation of a contradiction from the assumption of A
- a proof of $A \lor B$ is either a proof of A or a proof of B
- a proof of $A \land B$ is a proof of A and a proof of B
- a proof of $A \to B$ is a construction that transforms any proof of A into a proof of B
- a proof of $\forall x\ A$ is a construction that for any given object a yields a proof of $A(a)$
- a proof of $\exists x\ A$ is the specification of an object a and a proof of $A(a)$

This semantics entails that some classically valid laws must be abandoned. At the level of propositional logic, the law of excluded middle, $A \lor \neg A$, provides a famous example. On the BHK interpretation, this "law" states that every mathematical claim can either be proved or refuted, which is clearly unacceptable. So this interpretation opens up some space between A and $\neg A$, which is not available on the classical interpretation in terms of two truth-values. A closely related restriction concerns proof by *reductio ad absurdum*, which is no longer licensed. To prove a claim A by *reductio*, we assume its negation, $\neg A$, and try to show that a contradiction ensues. If we succeed, we have on the BHK interpretation established $\neg\neg A$. But this falls short of possessing a proof of A. Further examples of laws that are valid classically, but not intuitionistically, arise with the quantifiers. For instance, we must abandon the law $\neg\forall x\ A(x) \to \exists x\ \neg A(x)$. Even if it is contradictory to assume that we can prove $\forall x\ A(x)$, we need not be able to construct a counterexample, as the consequent asserts.[6]

Our discussion of intuitionistic logic has so far been purely technical. Let us now consider the philosophical question of why one might want to abandon classical logic in favor of intuitionistic. Different answers have been proposed, starting from different philosophical assumptions. Some of the answers converge on the BHK interpretation. One example is Brouwer's

---

[6] However, this law can be derived using intuitionistically valid rules for the quantifiers and *classical* propositional logic.

antirealist conception of mathematical truth, which holds that to be true is to be constructed or proved. The classical assumptions concerning truth are therefore flawed and must be replaced by valid assumptions concerning constructive proof.

Another example derives from the work of Michael Dummett (1925–2011) and is based on considerations about language and the possibility of objective communication. (The argument is thus unlikely to have appealed to the antilinguistic and subjectivistic Brouwer.) Dummett pitches a proof-conditional semantics against its traditional truth-conditional alternative, which regards the meaning (or "semantic value") of an expression as its contribution to the truth-value of the sentence. As Dummett observes, many of the resulting truth-conditions are *verification-transcendent*, in the sense that there is no effective procedure for determining whether the condition is satisfied. Examples include statements about the distant past, unmanifested dispositions, and especially infinite totalities. So far, so good. The most controversial part of Dummett's argument is the contention that verification-transcendent truth-conditions are problematic. This contention is defended by means of two challenges. According to the *acquisition challenge*, we could never learn such truth-conditions. Moreover, the *manifestation challenge* denies that we could ever manifest our understanding of such truth-conditions, which calls into question our claim to possess such understanding. Both challenges are controversial and have generated much debate. Since this is a debate in the philosophy of language, not specifically about mathematics, we shall not pursue it here.[7]

Let us instead zoom out a bit. A standard objection to intuitionistic logic is that it does violence to our existing inferential practice. The task of the philosopher of mathematics, the objector says, is to account for mathematical practice as we actually find it, not to revise it. Why "deprive the boxer of the use of his fists"? The objector is certainly right that philosophers should be wary of dictating to mathematicians (or any other scientists) how to go about their business. It takes a good reason to change what for all intents and purposes is a successful scientific practice.

---

[7] See Dummett (1978a) and, for a survey, Hale (1997).

Even so, it would be a mistake to insist that philosophers must always "leave everything as it is" (as Wittgenstein once put it) and limit themselves to a humble commentary on existing scientific methodology. There is ample room for critique of existing scientific practice within science itself, and such critique sometimes results in powerful reasons for change (cf. §1.5). Moreover, it is important to bear in mind that this critique may go beyond the concerns of any individual branch of science. In short: the intuitionists offer *reasons* for their rejection of classical logic, and each such reason needs to be assessed on its own merits (as we attempted to do in the previous section), not summarily dismissed on the grounds that it conflicts with existing mathematical practice.

5.4 INTUITIONISM AND POTENTIAL INFINITY

I turn now to a third route to intuitionistic logic. Unlike the previous two routes, this one avoids any appeal to the BHK interpretation or an antirealist identification of mathematical truth with our possession of a proof. Rather, the third route is based entirely on considerations about potential infinity.[8]

Consider the claim that the natural numbers are merely potentially infinite (cf. §4.4). The idea is that for any given natural number, it is possible to generate its successor, but that it is impossible to complete this process of generation. As we have seen, this idea is naturally explicated by the following two claims:

(5.1) $\Box \forall m \Diamond \exists n \, \text{SUCCESSOR}(m, n)$

(5.2) $\neg \Diamond \forall m \exists n \, \text{SUCCESSOR}(m, n)$

A strict defender of potential infinity may nevertheless deny that this goes far enough. The strings $\Box \forall m$ and $\Diamond \exists m$ behave much like the classical mathematician's quantifiers $\forall m$ and $\exists m$:

---

[8] See Linnebo and Shapiro (2016), which expands on the discussion in this section.

they serve as devices for generalizing over absolutely all natural numbers, not just the ones generated so far, but also all that could be generated in the future. But arguably, if we are serious about the mere potential infinity of the natural numbers, it is not enough to insist that every number be generated after finitely many steps. We must additionally require that every arithmetical truth be *made true* after finitely many steps. Since this takes us beyond the liberal form of potentialism outlined in §4.4, let us call the resulting view *strict potentialism*.

For simplicity, let us suppress the modal operators and revert to the usual quantifiers $\forall m$ and $\exists m$ to range over all natural numbers—albeit understood as merely potentially infinite. The question is whether these quantifiers can be made sense of from a strict potentialist point of view. The existential quantifier poses no special problem. An existential generalization $\exists m\, \varphi(m)$ is plausibly seen as made true by a true instance $\varphi(\bar{n})$, where the numeral $\bar{n}$ is generated after finitely many steps. But universal generalizations are problematic. A strict potentialist cannot allow a universal generalization to be true merely by virtue of the totality of all possible constructions of numbers; for there is no such totality. To be permissible, a universal generalization has to be made true after finitely many steps. But it is hard to see how this might be possible, given the formula's concern with infinitely many numbers.

We saw that Hilbert grappled with the same problem in his analysis of finitism (cf. §4.4). His answer was to understand universal generalizations over the natural numbers as schematic generalities. For example, the equation $a + 1 = 1 + a$, where a is a variable for natural numbers, can be understood schematically as the claim that for any numeral that might be produced, the corresponding substitution instance of the equation is true. Our knowledge of this schematic claim is based not on an ability to survey the natural numbers in their entirety but on an effective procedure for proving any instance we may confront; namely, given any numeral $\bar{n}$, we know how to reduce the expressions $\bar{n} + 1$ and $1 + \bar{n}$ to canonical numerals and observe that these are identical. But a serious problem remains, namely that a schematic generality is incapable of negation. For example, $a + 1 \neq 1 + a$

83

yields a universally generalized negation, not the desired negated universal generalization.

Fortunately, we can do better. The universal generalization $\forall m\, \varphi(m)$ can be seen as made true by the existence of an effective procedure which, when applied to any numeral $\bar{m}$ that might be generated, yields a "truth maker" for the corresponding instance $\varphi(\bar{m})$. Provided we require that the procedure be available (or generated) at some finite stage, this will satisfy the strict potentialist. For the universal generalization will then be made true at the finite stage where the procedure that acts as its "truth maker" becomes available. We do not, however, require that *we possess* such a procedure. So there is no reliance on antirealism.

Of course, this talk about effective procedures and "truth making" must be made precise. One attractive option is to invoke the *realizability interpretation* of intuitionistic logic. First, the notion of an effective procedure is explicated as in computability theory. Then, a formula is said to be true just in case there is an effective procedure that acts as its "truth maker"—or *realizer*, to use the customary technical term. Equipped with the resulting definition, we can ask whether we obtain "the right" truths. A natural measure of what is "right" is provided by the standard intuitionist theory of arithmetic, known as *Heyting arithmetic*, whose axioms are those of first-order Dedekind-Peano arithmetic (cf. §11.2) but whose logic is intuitionistic, not classical. It is a pleasing fact that every theorem of Heyting arithmetic is true according to our precise definition. Indeed, we have the following theorem.[9]

Theorem. Assume φ is a theorem of Heyting arithmetic. Then Heyting arithmetic (and thus also Dedekind-Peano arithmetic) proves that φ has a realizer. However, there are theorems of Dedekind-Peano arithmetic that do not have a realizer.

In sum, we have sketched an account of quantification over the natural numbers which satisfies the strict potentialist's requirement that every arithmetical truth be made true after some finite

[9] See, e.g., Troelstra (1998).

number of steps. The account validates all the theorems of intuitionistic arithmetic, but not classical. As stressed throughout, the locomotive of this argument for intuitionistic logic is the strict potentialist's insistence that each arithmetical truth be made true at some finite stage, not an antirealist conception of these truths as grounded in our possession of a proof.

5.5 INTUITIONISTIC REAL ANALYSIS

The standard intuitionistic and classical theories of first-order arithmetic share the same mathematical axioms and differ only concerning the logic that is available. More substantive differences between intuitionistic and classical mathematics emerge when we look beyond arithmetic to real analysis.

In classical mathematics, one of the ways to represent the real numbers is as convergent sequences of rational numbers.[10] On this conception, each individual real number involves the actual infinite, since the sequence representing it is actually infinite. So this approach to the real numbers is clearly not available to intuitionists, who reject the actual infinite. Brouwer developed an elegant and powerful alternative, which nicely illustrates my earlier claim that intuitionism isn't only about taking away parts of classical mathematics but also about giving back some beautiful new mathematics. The basic ideas behind Brouwer's approach are fairly simple. Let us begin with the notion of a *choice sequence*, which is a potentially infinite sequence of objects. The sequence is thus "forever in the making," not a completed list of items. There may be constraints on the choice of objects to be added to the sequence, but within the limits set by the constraints, all the choices are free and arbitrary. It is useful to imagine each choice sequence as associated with an immortal clerk, who is responsible for selecting objects, one per second, chosen randomly within the limits set by the constraints.

---

[10] More precisely, a real is represented as a *Cauchy sequence*, which is a sequence of rational numbers q_i, for $i \in \mathbb{N}$, such that for every $\epsilon > 0$ there is a number N such that $|x_i - x_j| < \epsilon$ provided $i, j \geq N$.

Brouwer's brilliant idea is to understand a real number as a choice sequence of rational numbers, subject to the constraint that the sequence be convergent.[11]

To convey the flavor of the real analysis that ensues, consider functions from reals to reals. In classical mathematics, such functions are understood as an infinite set of ordered pairs of real numbers, each pair specifying an argument and the value that the function takes for this argument. This understanding is obviously unavailable to intuitionists, who foreswear completed infinite collections. Brouwer proposes instead to understand a function of the desired sort as a procedure for computing the value of the function on the basis of information provided about the argument. Moreover, each such procedure needs to be understood in a way that is compatible with the rejection of actual infinities. In order to approximate the value of the function to some specified degree of precision, we must never require more than a finite amount of information about the argument; otherwise we would have to "wait until the end of time" when the argument has been fully specified. Let us make this entirely precise. Suppose we are interested in the value of a function f on an argument a, which is a choice sequence specifying a real number. Then, for any positive rational number ϵ, there must be some natural number N such that information about the N first terms of a suffices for the procedure that specifies f to determine the value $f(a)$ with an error of less than ϵ. It is not hard to see that this requirement implies that every function from reals to reals is continuous.

This theorem of intuitionistic real analysis conflicts with classical analysis, which makes extensive use of discontinuous functions. There is, for instance, a "step function" whose value is 0 for arguments less than π, and 1 thereafter. So the disagreement between intuitionists and classicists is far greater in analysis than in arithmetic. In arithmetic, the intuitionistic alternative is

---

[11] A sequence $\{x_i\}$ is *Cauchy convergent* just in case, for any $\epsilon > 0$, there is a number N such that $|x_i - x_j| < \epsilon$ whenever $i, j > N$.

strictly weaker than the classical one,[12] whereas in analysis, the intuitionistic alternative is sometimes weaker than the classical option *but other times contradicts it.*[13]

In light of its radicalism, it is unsurprising that intuitionistic mathematics has encountered objections. We already discussed the objection that the intuitionistic approach is too revisionary vis-à-vis accepted mathematical practice (cf. §5.3). A different, though related, objection is that intuitionistic mathematics is too weak for the purposes of empirical science. Whether this charge is correct is a matter of controversy.[14] But even if it is, this need not be fatal to intuitionism. As the great mathematician and sometime intuitionist Hermann Weyl (1885–1955) observed, an intuitionist may well permit classical mathematics to be used for the purposes of empirical science, where the certainty and apriority to which pure mathematics aspires is anyway unattainable.[15]

SELECTED FURTHER READINGS

Brouwer (1913) is a good introduction to intuitionism by its founder. Two more accessible presentations are given by his most famous student (Heyting, 1931, 1956). Dummett (1978a) present a much discussed attempt to revive a form of intuitionism. Two useful recent texts are the survey by Iemhoff (2015) and a short and accessible monograph by van Atten (2004) on Brouwer's philosophy and mathematics.

---

[12] Classical arithmetic can nevertheless be interpreted in intuitionistic arithmetic and is thus consistent provided that the latter is. See Moschovakis (2015, §4.1).

[13] These comparisons take the relevant languages at face value. If this assumption is rejected, the two approaches can be taken to have different subject matters, and the apparent disagreements, to be merely verbal. This affords the possibility of a pluralistic conception on which the two forms of mathematics live happily side by side.

[14] See McCarty (2005, IV.2) for references and a case that the charge is incorrect.

[15] See Weyl (1949, pp. 61–62).

CHAPTER SIX

Empiricism about Mathematics

6.1 Might Mathematics Be Empirical After All?

We have investigated a variety of attempts to make sense of mathematics. But all the attempts considered so far share Plato's conviction that mathematical knowledge is *a priori*. The apparent existence of *a priori* knowledge has always posed a problem for empiricists, who hold that all substantive knowledge is based on sense experience. Particularly offensive to empiricists is the view that mathematical knowledge is not only *a priori* but also synthetic, or nonconceptual. As we have seen, this view is held by Plato, Kant, and the intuitionists—although only Kant and perhaps the intuitionists went on to take the Copernican turn and insist that the objects of mathematical knowledge must conform to our faculty of intuition, rather than the other way round. Regardless of the details, the very idea that mathematical knowledge is synthetic *a priori* clashes with the empiricist creed that all substantive knowledge is empirical.

What, then, are empiricists to make of mathematics? From David Hume up through the logical empiricists of the early twentieth century, the favored response was to admit that mathematical knowledge is *a priori* but to deny that it is substantive enough to count as a counterexample to empiricism. Thus, Hume famously claimed that mathematical knowledge concerns "relations of ideas" rather than "matters of fact."[1] The game formalists go even further in their attempt to avoid substantive mathematical knowledge. By denying that mathematical language has any content at all, they reduce mathematics to a purely formal game.

---

[1] Although this anticipates Frege's view that arithmetic is analytic, Frege was certainly no empiricist. He even accepted that geometry gives us synthetic *a priori* knowledge.

There is a more direct response to the challenge that mathematics poses to empiricism, however. Instead of maneuvering to explain why *a priori* mathematical knowledge is no counterexample to empiricism, why not simply reject the old Platonic idea that mathematical knowledge is *a priori*? Surprisingly few empiricists have attempted this direct response. This chapter is about some of the brave exceptions. First out was John Stuart Mill (1806–73), who defended a brand of empiricism thoroughgoing enough to include even mathematics. A more sophisticated option was later developed by W. V. Quine (1908–2000), who first rejected the old Kantian dichotomies in terms of which philosophical questions about mathematics have traditionally been posed, and then put in their stead a relaxed and holistic form of empiricism. This Quinean strategy lives on today in the form of the so-called *indispensability argument*, which seeks to account for mathematics in terms of its indispensable contribution to empirical science.

6.2 MILL'S EMPIRICIST ACCOUNT OF ARITHMETIC

While our main focus will be on Quine and the indispensability argument, we begin with a brief exposition of Mill's radical empiricism and some problems it encounters. We shall focus on the case of arithmetic.

Frege famously held that numbers are ascribed to concepts, not to portions of physical matter (cf. §2.3). This allows him to say that the cards are 52, while the suits are only 4, although the physical matter involved in both number ascriptions is the same. Mill defended the competing view that numbers are ascribed to "aggregates." He realizes, however, that each "aggregate" needs to be articulated into distinct objects in order to handle examples such as the one just mentioned. Simplifying somewhat, it is useful to think of a Millian "aggregate" as just one or more things considered together—or a "plurality," as I shall often say for convenience.

The resulting plural discourse receives an attractive analysis in the form of *plural logic*. This logic supplements the ordinary 89 singular variables of first-order logic with special plural variables.[2] As usual, each singular variable has a unique value. But each plural variable is allowed to have one *or more* values. On this interpretation, plural logic does not incur any new commitments to objects such as classes but encodes irreducibly plural reasoning about the objects to which we were already committed. In particular, the many values of a plural variable are no less accessible to our senses than the single values of ordinary variables.

Mill's alternative to the Fregean analysis is that a numeral represents a cardinality property of some things.[3] For example, '2' represents the property that some things have just in case there are two of them, that is, just in case the things in question include one thing and another thing but no further things. It is important to notice that this cardinality property is *nondistributive*. While some things are cards just in case each of them is a card, it is not the case that some things are two just in case each of them is two. That is, *being a card* is distributive, while *being two* is not. Mill's view is thus that arithmetic is concerned with pluralities of things and their nondistributive cardinality properties. This account of the meaning of the numerals is entirely acceptable to an empiricist. The pluralities in question are just bunches of things, which may be assumed to be empirically accessible. And nondistributive properties are no more problematic, from an empiricist point of view, than distributive ones.

What about the laws of arithmetic? These are laws relating nondistributive cardinality properties, Mill claims. For example, $2 + 2 = 4$ states that the result of combining two disjoint pairs of things is a quadruple of things. The full extent of Mill's radicalism emerges only when we ask how the laws of arithmetic are known. As a thoroughgoing empiricist, Mill answers that the laws are known empirically by being inductively confirmed by their instances. For example, since every observed combination of two disjoint pairs has been a quadruple, we make the bold

<hr/>

[2] See Linnebo (2012) for an overview. The plural variables are often written xx, yy, etc.

[3] Strictly speaking, the numeral "connotes" such a property (or "attribute"), while it "denotes" the plurality (or "aggregate").

inductive generalization that these cardinality properties will always continue to be thus related.

Various problems with Mill's view were identified already by Frege.[4] To begin with, the view bears little resemblance to actual mathematical practice. It is true that children are taught arithmetic by being trained to count pluralities of things. But this is a matter of acquiring the relevant concepts, not of gathering evidence. Neither beginners nor experts regard a simple equation such as $2 + 2 = 4$ as better confirmed the more of its instances have been observed. Rather, once we have mastered the concepts, we can *prove* that this equation holds without exceptions—or so it seems.

Furthermore, we have direct empirical confirmation of only a minute fraction of all true arithmetical equations. So the pluralities we have actually counted are all tiny compared with infinitely many vastly larger uncounted pluralities. This means that our "data sample" is strongly biased in favor of small numbers. And this bias is ineliminable because of our finite capacities and the infinitude of numbers. How, then, can our empirical data inductively confirm the general laws of arithmetic, such as the axioms formulated by Dedekind and Peano? Why should these general laws continue to hold for truly large numbers, which for all we know may be importantly different from all the small numbers to which our data pertain?

Finally, any scientifically respectable use of empirical induction requires some statistics.[5] For instance, we need to know how many instances have to be observed in order for an experiment to be statistically significant. But statistics is a branch of mathematics, which stands in need of philosophical justification just as much as arithmetic does.

All of these problems point to a more general lesson. To require direct inductive confirmation of each arithmetical fact, let alone of mathematical facts more generally is hopeless.

---

[4] See Frege (1953, esp. §§7–10 and 23–24). We set aside some of Frege's weaker objections that are answered simply by clarifying, as we have just done, the notions of an "aggregate" and a nondistributive property.

[5] See, e.g., Frege (1953, fn. 1 and §10).

Any viable form of empiricism about mathematics has to allow a theory to be confirmed as a whole and mathematical facts to be confirmed only indirectly via their contributions to this theory. This lesson takes us to our next great empiricist, who is a staunch defender of this form of holistic confirmation.

6.3 QUINE'S HOLISTIC EMPIRICISM

Quine's empiricism differs sharply from Mill's. There is no need to accommodate mathematics within the category of synthetic *a posteriori* knowledge, Quine thinks. It is better to reject the analytic-synthetic dichotomy altogether and use the extra elbow room that ensues to develop a more relaxed and holistic form of empiricism that includes mathematics.

The attack on the analytic-synthetic distinction is launched in one of the greatest classics of twentieth-century philosophy (Quine, 1953a). The article opens by criticizing various attempts to define or explain the distinction. A good example is Frege's famous definition of a sentence as analytic just in case it can be transformed into a logical truth by replacing expressions with synonymous expressions. For instance, "All bachelors are unmarried" counts as analytic, since it can be transformed to the logical truth "All unmarried men are unmarried" by replacing 'bachelor' with the synonym 'unmarried man.' But Quine objects that Frege's definition of analyticity offers little progress, as it relies on the equally problematic notion of synonymy.

The most promising definition, according to Quine, is based on a *verificationist* account of meaning, which takes the meaning of a sentence to be characterized by the collection of (actual and possible) sensory observations that would confirm the sentence. On this account, a sentence can be defined as analytic just in case it will be confirmed "come what may," that is, by any possible sensory observation. This definition is meant to improve on Frege's by eschewing any notion of meaning, which Quine takes to stand in equally great need of explanation. Even so, Quine finds fault with the attempted definition. It ignores the important phenomenon of *confirmational holism*, namely that no sentence is confirmed in isolation but only as part of a larger body of beliefs. Most sentences are only indirectly concerned with observation and have consequences concerning the observable only when combined with other sentences. For example, the universal law of gravitation does not entail that two particles attract one another until we add that both particles have non-zero mass. In fact, even sentences that *are* directly concerned with observation can be held true "come what may" by making appropriate changes elsewhere in our "web of beliefs" so as to dismiss a potentially falsifying observation as a hallucination. The upshot is that, even if a verificationist account of meaning is assumed, confirmational holism makes it impossible to give an informative definition of analyticity.

Quine's attack on the analytic-synthetic distinction has generated an enormous literature, which we cannot survey here. Let us rather proceed directly to a powerful alternative conception of science that Quine proposes, which has no need to distinguish between the analytic and the synthetic. The alternative conception proposes that our beliefs be regarded as a vast and highly interconnected field. Observation impinges on the field only at its periphery, where we find sentences directly concerned with observational matters. Throughout the field, beliefs are connected by the logical relations in which they stand. This conception incorporates two central tenets of Quine's view. One is his *empiricism*. The only constraints on scientific theory are the sensory observations that impinge on the field's periphery. There are no analytic truths that serve as constraints. Nor is the field constrained by any rational insight (as in Plato) or any other nonempirical form of evidence (such as Kant's "pure intuition"). A second tenet is Quine's *confirmational holism*, which states that only science as a whole faces "the tribunal of experience." When the web of beliefs entails an observational prediction that is confirmed or disconfirmed, the resulting epistemic praise or blame cannot be assigned to any proper part of the web but accrues to it only as a whole. In this way, the empirical evidence received at the periphery permeates the entire the field, all the way to its core.

The resulting holistic empiricism has strong implications concerning mathematics. As for the metaphysics of mathematics,   93

Quine agrees with Frege that mathematics is concerned with abstract objects. After all, we have good reason to take mathematical theorems to be true, and these theorems are, by Quine's own criterion, ontologically committed to abstract objects. But Quine places greater emphasis on "ontological economy" than Frege did: he seeks to avoid postulating more objects than strictly necessary in order to do good science. This concern leads Quine to endorse a form of set-theoretic reductionism, which reduces all other mathematics objects to sets (cf. §3.5).

The true novelty of Quine's view emerges only when we turn to epistemology. Quine attempts to render mathematics scientifically respectable by assimilating it to the theoretical parts of empirical science. Mathematics, he claims, isn't essentially different from theoretical physics. Both go beyond what can be observed by means of our unaided senses. And both are justified by their contribution to the prediction and explanation of states of affairs that can be thus observed. As Quine puts it, "the objects of pure mathematics and theoretical physics are epistemologically on a par. ... Epistemologically the primary cleavage is between these on the one hand and observables on the other" (1986, p. 402).

Why, then, are we so reluctant to revise our mathematical beliefs? The only reason is their location at the core of our field of beliefs, Quine claims. This location means that a revision of our mathematical beliefs would have repercussions throughout the entire field. It is therefore pragmatically inadvisable to revise a well-established mathematical belief unless the reasons for doing so are extremely strong. Like any other part of the field, however, our mathematical beliefs are in principle open to revision.

6.4 PROBLEMS WITH QUINE'S HOLISTIC EMPIRICISM

Quine's empiricist account of mathematics is a step forward compared with Mill's. There is no need for mathematical truths to be confirmed in isolation. Mathematical theories can instead be confirmed holistically by virtue of their indispensable contribution to our overall science. But despite this progress, problems remain. We shall consider three.

According to Charles Parsons, Quine is unable to account for the obviousness of much of elementary mathematics.[6] Basic arithmetical beliefs seem to enjoy a far more direct and compelling form of justification than the highly indirect and cumulative evidence that is marshalled in support of our beliefs about theoretical physics. Our belief that $2 + 2 = 4$, for example, seems to differ radically in epistemological status from our belief that neutrinos have mass. But as we have seen, Quine insists that the two beliefs are "epistemologically on a par." Who is right? Things certainly *seem* to be as Parsons says. To be fully convincing, however, this observation must be supplemented with a worked-out account of the direct and compelling evidence that elementary mathematics allegedly enjoys. Parsons attempts to do so by developing a broadly Kantian account of mathematical intuition. Another option is to appeal to evidence flowing from a Fregean form of abstraction.[7]

A second problem with Quine's approach is that it leaves large parts of mathematics without scientific justification. The problem stems from his insistence that mathematical axioms earn their keep by contributing to empirical science. A mathematical axiom that fails to contribute in this way lacks scientific justification. As Quine puts it, his view of pure mathematics is "oriented strictly to application in empirical science" (1986, p. 400). The question is thus how much abstract mathematics we really need for the purposes of empirical science. One of the more accomplished mathematicians to have investigated the matter is Feferman, who contends that empirical science can live happily on a surprisingly meager diet, which leaves out vast parts of contemporary mathematics.[8] Certainly, only a tiny fragment of contemporary set theory is needed for the purposes of empirical science.

Quine's response to these considerations is two-pronged. On the one hand, he reminds us that ontological economy is not the sole consideration in theory choice; systematicity and

---

[6] See, e.g., Parsons (1980, §III).

[7] These two options will be discussed in Chapters 8 and 9, respectively.

[8] See, e.g., Feferman (1993).

elegance also matter. This might motivate some "rounding off" concerning the mathematics that we accept.[9] On the other hand, Quine admits that some widely accepted mathematical axioms may lack empirical justification, even of the indirect form. His verdict in such cases is uncompromising: such axioms must be regarded as "[m]athematical recreation ... without ontological rights" (ibid.). Given his empiricism, it is hard to see how he could have responded differently. However, by holding mathematics hostage in this way to the needs of empirical science, Quine contradicts prevailing mathematical methodology. Mathematical axioms tend to be assessed on their own merits, based on *intramathematical* considerations. In particular, the concern with ontological parsimony that goes hand in hand with Quine's holistic empiricism appears to play no role whatsoever in contemporary mathematics.[10]

A third problem is that mathematical theories never face "the tribunal of experience" in the way that empirical theories do. Rather, mathematical theories play a *constitutive role* in providing a framework without which the empirical theories could not even be formulated. A nice example is due to Michael Friedman, who argues that, when the theory of relativity was subjected to empirical tests, the underlying mathematical theory of Riemannian geometry never faced "the tribunal of experience" but was presupposed in order to formulate the physical theories that do. According to Friedman, this points to a deep asymmetry between the empirical part of the theory of relativity and the mathematics that this empirical theory employs, where Quine, by contrast, regards the two as on a par.[11]

Again, the three problems point to a more general lesson, namely that mathematics differs from the more theoretical parts of empirical science in its epistemic status and the nature of its contribution to science as a whole.

---

[9] For the set-theoretic *cognoscenti*: Quine (1986) even considers adopting $V=L$ in this spirit.

[10] See Maddy (1997).

[11] See, e.g., Friedman (1997) but also Putnam (1975a).

6.5 THE INDISPENSABILITY ARGUMENT

Despite these problems, there can be no doubt that Quine has exercised a strong influence on the philosophy of mathematics, especially in North America. Today this influence is most clearly manifested in *the indispensability argument*, which takes its departure from Quine and perhaps also Putnam.[12] The argument defends mathematics—understood in a broadly platonistic manner—by appealing to its indispensable contribution to empirical science. Let us begin by setting out the argument.

Premise 1. Existential quantifiers incur ontological commitment. That is, for a statement of the form $\exists x\, \varphi(x)$ to be true, there must be some object that satisfies the condition φ.

Premise 2. Natural science makes indispensable use of theories that quantify over abstract mathematical objects.

Premise 3. We have reason to believe what natural science tells us.

Conclusion. We have reason to believe that there are abstract mathematical objects.

The reasoning is straightforward and compelling. By Premise 2, theories that quantify over abstract mathematical objects are part of the natural science which, by Premise 3, we have reason to believe. Since these theories by Premise 1 incur ontological commitment to abstract objects, we have reason to believe that such objects exist. Thus, the success of the argument turns on the tenability of the premises. Let us take a closer look.

The first premise is Quine's theory of ontological commitment: "To be is to be the value of a bound variable."

---

[12] See the works by Quine cited in the previous section, as well as Putnam (1971) (although Putnam (2012, chap. 9) disowns any responsibility for the indispensability argument).

This premise is closely related to the thesis of Classical Semantics that we discussed in connection with Frege's argument that numbers are objects (cf. §2.3). So its tenability turns on the same considerations that come up in connection with that thesis.[13]

The second premise seems plausible. Even a glance at scientific publications suffices to reveal a heavy reliance on mathematics. But the premise has in fact proved controversial. Let us begin by clarifying it. It is plausible to understand the premise as stating that our best current scientific theory of the world uses mathematics that quantifies over abstract objects, and that this use cannot be eliminated without thereby compromising the scientific merits of the resulting theory. This precise statement of the premise suggests a nominalist response. We can try to develop an alternative but equally good scientific theory that eschews all qualification over mathematical objects. This response corresponds to one of the two main forms of nominalism that we discuss in the next chapter (cf. §§7.2–7.4).

The final premise embodies a form of *naturalism*, by which I mean an orientation that seeks to reduce, or even eliminate, the gap between philosophy and natural science. The premise states that well-established claims of natural science should be believed, not be trumped by distinctively philosophical considerations. This claim must be distinguished from a far stronger and more controversial form of naturalism based on the converse claim that the natural sciences offer a *complete* picture of the world and of our relation to it. The indispensability argument has no need for this stronger form of naturalism.

Even so, the third premise has encountered opposition. Why should we believe *all* the pronouncements of natural science? Even if we grant that natural science is *generally* a good guide to truth, might not *some* of its widely accepted claims lack justification? Not all the parts of a machine contribute in the same way. A machine may even continue to function despite a broken part. In short: we need to be told why someone who

---

[13] In §11.2 we consider eliminative forms of structuralism which deny Premise 1 as applied to the language of arithmetic.

has general faith in natural science should believe the claims of mathematics in particular. One response is to reintroduce Quinean confirmational holism, which the argument has so far avoided. Natural science is justified as a whole, and this justification is transmitted to any indispensable part of it. As we saw in the previous section, however, confirmational holism is controversial. Moreover, some nominalists argue that the use of platonistic mathematics in extant science is merely instrumental. And statements that are used in a purely instrumental way need not be true for science as a whole to succeed. This response corresponds to a second form of nominalism, to which we shall return (cf. §7.5).

I wish to end with a more general concern about the indispensability argument. Even if some version of the argument can be made to work, it would not follow that the argument provides our only—or best—reason to believe the claims of mathematics. Some reasons for doubt surfaced in our discussion of Quine's holistic empiricism. Whatever indirect support mathematics might obtain from its contributions to empirical science looks very different from the kinds of consideration that are operative in actual mathematical practice, which pays far less attention to the usefulness of mathematics in the empirical science. The problem is particularly serious for aspiring naturalists, who wish to respect successful sciences. As we observed at the beginning of the book, mathematics is a successful science *par excellence*. It is problematic to set aside methodological norms that are operative in this successful science in order to accommodate an empiricist conception of knowledge.[14] Beginning in Chapter 8, we shall therefore investigate some proposed nonempirical forms of evidence in mathematics.

SELECTED FURTHER READINGS

Skorupski (1989, chap. 5) discusses Mill's view of mathematics. Quine's view is developed in several work (1953b, esp. §§5–6;

---

[14] See Maddy (1997).

1986; 1992, §40; and 1995, chap. 5.) (Colyvan (2015 provides a useful survey of the indispensability argument. Putnam (1971, chap. 8), is widely regarded as an early statement of the argument. Maddy (1992) is an influential critique of the argument. Baker (2005) attempts to develop an "enhanced" indispensability argument, which focuses on the alleged indispensable *explanatory* role of mathematics in science.

Nominalism

7.1 BENACERRAF'S DILEMMA

In contemporary philosophy of mathematics, 'nominalism' typically refers to the view that there are no abstract objects. The *locus classical* for the debate about nominalism in this sense is Benacerraf's "Mathematical Truth" (1973).[1]

Benacerraf's central contention is that there is a conflict between two desiderata. On the one hand, we want *a plausible semantics for the language of mathematics*. Ever since Frege, the default option has been what we have called "classical semantics," according to which the semantic function of singular terms is to refer, and that of quantifiers, to range over appropriate domains of entities (cf. §2.3). This semantics provides a unified account of mathematical and nonmathematical language, which is widely felt to be attractive. After all, sentences whose syntactic structure is the same should presumably receive the same semantic analysis, regardless of what the sentences are about. As Frege observed, however, when classical semantics is applied to the language of mathematics, this naturally leads to the postulation of mathematical objects. For example, if "$2 + 2 = 4$" is true and its singular terms function as described in classical semantics, there must be objects denoted by the numerals '2' and '4.' This is just an instance of the Fregean argument for *object realism*, which we defined as the view that there exist mathematical objects. And presumably, these objects are abstract.

On the other hand, we want *a plausible epistemology for mathematics*. As Benacerraf put it, "[i]t must be possible to link

---

[1] Throughout most of the history of philosophy, 'nominalism' referred to the different view that there are no universals (i.e., properties that can be instantiated by many different particulars), only particulars.

up what it is for *p* to be true with my belief that *p*" (1973, p. 667). Without such a "link," it would be a cosmic accident that my methods for forming mathematical beliefs result in beliefs that are true, at least for the most part. As we have seen, the apparent need for such a link has been discussed ever since Plato's *Meno* and corresponds to what we have called "the integration challenge" (cf. §1.5). But Benacerraf goes beyond his predecessors by taking far greater care to rule out fanciful accounts, such as Plato's speculations about prenatal learning or the rationalists' postulation of a faculty of reason alleged to provide the requisite link. Benacerraf requires a naturalistic—or broadly scientific—account of the link. And such an account, he believes, has to be based on a *causal connection* between the agent in question and what is known.

As Benacerraf observes, there is a conflict between the two desiderata, at least as he understands them. The first desideratum leads to an account on which mathematics is concerned with an ontology of abstract objects. But the second desideratum requires a causal connection between the knower and the known. Since abstract objects by definition don't participate in causal relations, no such connection is possible. What to do?

Benacerraf remained undecided about which desideratum to relinquish, but others have expressed strong views. One family of responses to Benacerraf's dilemma insists on a naturalistically acceptable epistemology of mathematics and contends that this demand can be met only if we avoid all commitment to abstract objects. How can this avoidance be achieved? Here it is useful to recall the Fregean argument for the existence of abstract objects (cf. §2.3). Each of the argument's three premises can be denied. One option is to deny that there is any substantive role for *truth* in mathematics. This was done by the game formalists and was recently tried by Hartry Field, as we shall see shortly. Another option is to deny that the language of mathematics should be given a *classical semantic analysis*. This was done by the term formalists and, more recently, by a modal form of structuralism to be discussed in §11.2. Last, one may deny that mathematical objects are *abstract*. An interesting attempt to do so was made by Maddy (1990), who argues that sets of concrete objects inherit spatiotemporal location and causal efficacy from their elements (cf. §8.4).

Another family of responses to Benacerraf rejects the second horn of the dilemma, at least when this is taken to impose a causal constraint on knowledge. This too can be done in different ways. One option is to embrace confirmational holism. If scientific theories are confirmed only as a whole, not statement by statement, this lessens the need for a direct "link" between a knowledgeable belief and the subject matter of this knowledge. The viability of this approach was discussed in the previous chapter. Another option—which I and many others prefer— is to deny that all forms of knowledge are subject to a causal constraint. Let us therefore examine why Benacerraf insists on such a constraint.

Benacerraf is, as we saw, keen to ensure that the integration challenge be answered in a scientifically responsible manner. What does this mean, however? A natural strategy is to examine cases of knowledge where the challenge can be met, such as knowledge based on perception, testimony, and memory. In all these cases, an appropriate causal relation appears to be required. Might this suggest that *all* forms of knowledge require such a relation? This would be a terrible argument, however! As we have seen, mathematics is a very successful science, which is also strikingly different from the paradigmatic empirical sciences. The proposed generalization would therefore be problematic. The defender of abstract objects could simply counter that it is a rash *over*generalization, to which mathematics provides a counterexample. Unlike the integration challenge, which is carefully formulated so as *not* to be biased against any particular form of knowledge, the demand for a causal connection *is* biased.

Another possible source of inspiration for Benacerraf's causal requirement on knowledge comes from epistemology. Traditionally, knowledge was often defined as justified true belief. Some famous counterexamples showed this definition to be untenable.[2] This discovery sparked the search for a better definition. Many philosophers sought to add another condition—often thought to

---

[2] See Gettier (1963).

involve causality—to the traditional definition. One influential example is Alvin Goldman's causal theory of knowledge, to which Benacerraf appeals.[3] Since then, however, epistemologists have largely abandoned the causal theory of knowledge. Moreover, in a clash between a philosophical theory of knowledge and the successful science of mathematics, it seems foolhardy to side with the former. Mathematics commands far greater confidence than any philosophical analysis of knowledge.[4]

To conclude, it has not been established that knowledge requires a causal connection between the knower and the known. This undermines the influential epistemological objection to object realism about mathematics. Yet it would be premature for object realists to declare victory. It is one thing to reject an unreasonable demand on knowledge and quite another to provide a positive account of mathematical knowledge that allows the integration challenge to be met. Object realists still face a formidable challenge. And there are other challenges as well.[5] So it remains interesting to examine the prospects for a nominalistic account of mathematics. This will occupy us in the remainder of this chapter. In the chapters that follow, we shall change tack and explore how object realists can attempt to answer the integration challenge.

7.2 HARTRY FIELD'S STRATEGY FOR NOMINALIZING SCIENCE

According to Field, "the *only* serious argument for platonism depends on the fact that mathematics is applied outside of mathematics" (1989, p. 8). In order to undermine this single serious argument, he sets out to show how nominalists too can explain applications of mathematics to the empirical sciences.[6]

---

[3] See Goldman (1967). However, Goldman states that his only concern is empirical knowledge (p. 357).

[4] See, e.g., Lewis (1986, pp. 108–15).

[5] See, e.g., Linnebo (2013a, §4) for an overview and references.

[6] See Field (1982) for an accessible overview and Field (1980) for the full account.

If successful, this explanation will show that mathematics is not, after all, indispensable to empirical science.

Field's argumentative strategy is inspired by a toy example due to Putnam, discussed in connection with game formalism (cf. §3.2). We observed that first-order logic provides an ontologically innocent way to make finite number ascriptions. For example, the claim $\#x\, Fx = 1$ (which is "platonistic" because of its reference to the number 1) can be "nominalized" as $\exists x \forall y (Fy \leftrightarrow x = y)$.[7] We also observed, however, that it is exceedingly impractical to work with such nominalistic number ascriptions, as the formulas and derivations quickly become too long to be surveyable. It is hugely advantageous to allow the claims of ordinary platonistic arithmetic too, as well as bridge principles that link these claims with the nominalistically acceptable number ascriptions. For example, if there is a single F, a single G, and nothing that is both F and G, we can use the equation $1 + 1 = 2$ to infer that there are precisely two things that are F-or-G. This means using a detour through discourse about abstract objects such as numbers to simplify our reasoning about the concrete. Are such detours safe? Do we know that they won't take us from true nominalistic premises to a false nominalistic conclusion? In the present example, we do. We can prove that everything that can be established via a platonistic detour can also be established directly, remaining strictly within the limits of what is nominalistically acceptable.[8] In technical parlance, we can prove that platonistic arithmetic is *conservative* over the mentioned part of nominalistic arithmetic.

In light of this toy example, it is easy to explain the gist of Field's strategy for nominalizing science. The idea is simply to extend the strategy from the toy example to science in general. This dauntingly ambitious aim involves two separate tasks. First, we need to do to every scientific theory what we did to finite number

[7] In the remainder of the chapter, we shall follow Field in referring to any claim or theory committed to abstract objects as "platonistic," although this convention is somewhat less demanding than our official definition from §1.4.

[8] *Proof sketch.* Since finite cardinality facts are expressible in first-order logic and the cardinality entailments in question hold in all models, the entailments are also provable by the completeness of first-order logic.

ascriptions, namely to "nominalize" the theory by reformulating it in a way that avoids all commitment to abstract objects. Second, we need to show that the detours via platonistic mathematics are benign and serve only to simplify reasoning that could in principle be conducted while remaining at the nominalistic level. That is, we need to show that the platonistic theory is conservative over the nominalistic one. Suppose Field is right that the two tasks can be carried out. Then science can in principle be done in a nominalistic way (by the first task). It might nevertheless be expedient to apply mathematics to science as a device that is useful for practical purposes, although in principle eliminable (by the second task). In short, a successful execution of the strategy would yield an elegant and powerful account of "the fact that mathematics is applied outside of mathematics," just as Field desires.

7.3 CAN SCIENCE BE NOMINALIZED?

Let us begin by examining the first task. It needs to be made plausible that every scientific theory can be rewritten to eliminate all reference to numbers and other abstract objects in favor of statements solely about the physical world. Field attempts to make a good start by carrying out the task in the special case of the Newtonian theory of gravitation. While this theory is much simpler than today's best physical theories, its nominalization is far from straightforward and illustrates many of the general challenges that the task confronts.

In geometry it is traditional to distinguish between coordinate-free (or *synthetic*) and coordinate-based (or *analytic*) approaches. Ancient geometry is synthetic: it talks about points, lines, and planes, but not about numbers. Thanks largely to Descartes, however, the analytic approach came to dominate. Each point is now associated with a triple of real numbers, known as its *coordinates*. Geometrical objects can thus be described and investigated indirectly via their coordinates, which fall within the purview of powerful algebraic techniques. This development led to the differential calculus, which nicely illustrates the power and elegance of the analytic approach.

When nominalists describe physical space, they have good reason to prefer the synthetic approach, as this eschews numbers in favor of points and regions that can arguably be understood as physical. Thankfully, the synthetic approach too has received sophisticated modern developments, in particular by Hilbert and Tarski. Both use two predicates defined on points: $B(x, y, z)$ for "y lies between x and z" and $D(x, y, z, w)$ for "x is as distant from y as z is from w." Using these predicates, axioms for three-dimensional Euclidean geometry can be stated. How does the resulting synthetic geometry relate to the more familiar analytic geometry based on the set $\mathbb{R}^3$ of triples of reals? The question receives a satisfying answer by a *representation theorem*, which states that the two geometrical theories agree, in the following precise sense.[9]

Theorem. There is an isomorphism f from the set E of spatial points to the set $\mathbb{R}^3$ of coordinates. Thus, f is one-to-one and onto, and for every x, y, z, and w

$$B(x, y, z) \leftrightarrow B^*(f(x), f(y), f(z))$$

$$D(x, y, z, w) \leftrightarrow D^*(f(x), f(y), f(z), f(w))$$

where B^* and D^* are standard analytical formulations of the relevant claims about betweenness and distance, respectively.

As Field explains, the synthetic approach is easily extended to four-dimensional Newtonian spacetime, where an analogous representation theorem can be proved.

Next in line are quantities, such as mass and charge. Consider mass. It is customary to choose some massive object as a unit, say the standard kilogram in Paris. The mass of any other object can now be specified by comparing it with that of our unit. An object is said to have mass 2 kilograms, for example, just in case it is twice as massive as our unit. To make things precise, let '$x \preceq y$' express that x is no more massive than y, and let '$S(x, y, z)$' express that z is as massive as x and y taken together. Using these

---

[9] See Field (1980, p. 50) for a precise statement. The theorem can be proved by pooling the synthetic and analytic axioms.

predicates, we can formulate some plausible assumptions concerning physical objects that allow us to prove another representation theorem, which loosely speaking says that the family of mass properties and the operation of concatenation have a structure that is "mirrored" in that of the nonnegative real numbers and the operation of addition. The latter can thus be used to represent the former. More precisely, there is a function m from physical objects to nonnegative reals such that

$$x \preceq y \leftrightarrow m(x) \le m(y)$$
$$S(x, y, z) \leftrightarrow m(x) + m(y) = m(z)$$

where $m(x)$ is the mass of an object x in terms of our arbitrarily chosen unit, which m maps to 1.[10]

Representation theorems such as those just mentioned provide valuable insight into how numbers are relevant to physical space and to the various properties of the objects that populate this space. The theorems also enable us to explain how Field's nominalization project proceeds. He develops synthetic theories not only of space and mass but of any other quantity invoked in the theory of Newtonian gravitation. Moreover, Field shows how other relevant notions, such as that of differentiation, can be defined synthetically, with no reliance on numbers. All this requires hard work.

Does Field succeed, then, in his first task? I shall briefly describe four lines of criticism. First, Field quantifies over points and arbitrary regions of physical space. As he observes, this assumes a *substantivalist* view of space according to which these geometrical objects really exist. This assumption is controversial in the philosophy of physics. A related assumption of Field's is even more controversial, namely that there are *completely arbitrary* regions; that is, that every set of real coordinates defines a region. It is far from obvious that physical space has the immensely rich structure that is realized by $\mathbb{R}^4$ and all of its arbitrary subsets.[11]

---

[10] See Krantz et al. (1971, p. 74).
[11] See, e.g., Maddy (1997, pp. 143–52).

Second, there may be branches of physics that cannot be nominalized in Field's way. David Malament claims there are theories on which Field's strategy seems not have a chance. One example is the appeal to all possible dynamical states of a system in classical mechanics. Another is the pervasive use of infinite-dimensional vector spaces (so-called Hilbert spaces) in nonrelativistic quantum mechanics.[12]

Third, some philosophers see a conflict between Field's nominalism and the naturalism that motivates his view. As we have noted, naturalism seeks to minimize the difference between philosophy and natural science; in particular, philosophers should always respect successful science unless compelling reasons to do otherwise arise within science itself. Now, physics is surely a successful science. As currently practiced, however, physics is not nominalistic but on the contrary is awash in reference to numbers and sets. Field is therefore committed to revising a successful science. Its "official" version must be rewritten to eliminate these offending forms of reference. Are his reasons for revision sufficiently weighty? And do the reasons arise within science itself, or can they be dismissed as illegitimate philosophical interference?[13]

A final complaint concerns the status of ordinary scientific claims involving mathematics. Field regards most such claims as false. Why, then, is our practice of making such claims so successful? As Yablo (2005) observes, there is "something strangely halfway" about Field's project. The project leaves open the question of how mathematics—which in ordinary science isn't just a superstructure erected on top of some nominalistic base—manages to be so useful without being true. Field is more concerned with the dispensability of mathematics in principle than its successful applicability in science as we find it. This raises a hard question. Suppose we manage to explain the applicability of mathematics in extant science without assuming its literal truth. Why should we then be so concerned about its dispensability in principle? We shall return to this question in §7.5.

---

[12] See Malament (1982, pp. 533–34).
[13] This line of criticism is developed in Burgess and Rosen (1997) and Maddy (1997).

7.4 IS MATHEMATICS CONSERVATIVE OVER NOMINALISTIC SCIENCE?

Field's second task is to show that mathematics is conservative over nominalistic science; that is, that any argument from nominalistic premises to a nominalistic conclusion that makes a detour via mathematics can also be given without such a detour. This rouses two questions. Is the conservativeness claim true? And if so, are the resources needed to establish the claim nominalistically acceptable?

Let us begin by giving a precise formulation of the conservativeness claim. Let P (for "platonistic") be a theory that is committed to abstract mathematical objects, and let N be a nominalistic theory. Then P is said to be *conservative over N* just in case:

> For any sentence A in the language of N, if $P + N$ implies A, then N implies A.[14]

However, logical implication can be defined either proof-theoretically or semantically. The two definitions are equivalent in first-order logic, which is complete.[15] But in second-order logic, which is incomplete, we obtain either a proof-theoretic or a semantic notion of conservativeness depending on which definition of implication we adopt.

Field believes there are general reasons to take mathematics to be conservative—in both senses—over nominalistic science.[16] One may worry that the claim is too sweeping to be true. For example, let N be second-order logic and P be extant mathematical physics. Clearly, P is not conservative over N, since P proves many nonlogical truths. However, this apparent

---

[14] This slightly simplified definition is equivalent to Field's on the benign assumption that all quantifiers in nominalistic statements are restricted to nonmathematical objects.

[15] A logic is said to be *complete* just in case it has a complete proof procedure; that is, just in case, when some premises semantically imply a conclusion, there is a formal proof of this conclusion from these premises. While first-order logic is complete, second-order logic is not. See, e.g., Boolos (2007).

[16] See Field (1980, chap. 1).

counterexample is easily dismissed. Field presumably intends his general claim to be restricted to theories P that consist only of pure mathematics and bridge principles that link nominalistic claims with mathematical ones. But a worry still remains. While we have encountered various examples of bridge principles, Field provides no general definition. Until a definition is in place, Field's general conservativeness claim lacks precise content. We shall therefore focus on some more restricted conservativeness claims where P consists of pure mathematics and bridge principles that Field in fact discusses.

Even these restricted conservativeness claims face a major stumbling block in the form of Gödel's incompleteness theorems (cf. §4.6). (This discussion involves some subtle technical arguments. Readers less interested in such matters may wish to skip ahead to the next section.) The key is to observe that arithmetical claims can be expressed in the language of Field's synthetic geometry. First we choose an interval to represent the number 1. Then we let multiples of this interval represent the other natural numbers; for example, two adjacent copies of the interval yield a new interval representing the number 2. Using this representation, we can express a Gödel sentence G for the nominalistic theory N. As we know from the first incompleteness theorem, N cannot prove G. Consider now the platonistic theory P that results from adding set theory to N. As Stewart Shapiro observes, P enables us to prove G.[17] Thus, we have a counterexample to Field's conservativeness claim: P can prove something that N cannot, namely G.

How might Field respond? One option is to exploit the fact that his nominalistic theory is second-order, which, as we have seen, means that proof-theoretic and semantic conservativeness come apart. While the above argument shows a failure of *proof-theoretic* conservativeness, it poses no threat to the *semantic*

---

[17] This last fact assumes that we include among our bridge principles the claim that every set of points defines a region. When we remove this claim, conservativeness can be proved after all, but instead we become unable to prove a representation theorem that Field's account requires. See Shapiro (1983) and Burgess (1984, §4) for precise statements and proofs.

conservativeness of mathematics over nominalistic science. So Field may hold on to the latter as his official conservativeness claim.[18] Another option is to cut the mathematical theory down to size. The nominalistic Gödel sentence G has never played a role in any established physical theory. It is thus an option to formulate a weaker mathematical theory which suffices for the purposes of mathematical physics, but which (unlike set theory) *is* proof-theoretically conservative over nominalistic science. John Burgess has described a candidate for such a theory.[19]

I shall end with a brief remark on the resources needed to prove the desired conservativeness claims. For Field's nominalism to be a stable position and not merely serve as a refutation of platonism, these resources had better be nominalistically acceptable. In this connection it matters greatly how our nominalists choose to respond to the challenge from the incompleteness theorem. The study of second-order semantic implication requires far stronger mathematical resources than the study of proof-theoretic implication. Field outlined a modal account of the notion of logical implication that is meant to enable a nominalistically acceptable investigation of the conservativeness claims. Suffice it to say that the nominalistic acceptability of this account remains controversial, especially as applied to second-order semantic implication.[20]

7.5 MATHEMATICAL OBJECTS AS REPRESENTATIONAL AIDS

Field emphasizes the *deductive* usefulness of mathematics, as applied to nominalistic base theories. Mathematics greatly facilitates the derivation of nominalistic conclusions from nominalistic premises. More recently, philosophers such as Joseph Melia and Stephen Yablo have placed greater emphasis on the *representational* usefulness of mathematics. Mathematics allows

---

[18] See Field (1985). The threat is avoided because the incompleteness theorems apply only to proof-theoretic implication.

[19] See Burgess (1984), which draws on work by Tarski and Kripke.

[20] See Field (1991) for the account and Shapiro (1993) for criticism.

us to express—often in simple and elegant ways—various nonmathematical claims that would otherwise have been difficult or downright impossible to express.

Here are some examples that nicely illustrate the idea.[21]

(3) There are 2.5 planets per star.

This claim can be true in infinitely many different ways, each of which is nominalistically expressible: there are five planets and two stars, or there are ten planets and four stars, etc. So when an astronomer asserts (3), it is reasonable to assume that the assertion is intended to be entirely about the physical world and thus nominalistically acceptable. True, the astronomer invokes the number 2.5. But she does not intend to make any claim about abstract objects such as numbers. Her intention is merely to express in a simple and elegant way the mentioned nominalistic content. Similar examples abound. Consider for instance the claim that the escape velocity from a body of radius R and mass M is $2GM/R$, where G is the gravitational constant. Here we quantify over numbers not because of any intrinsic interest in them but merely as a representational aid that enables a compact and elegant expression of a family of purely physical facts.

These examples suggest a daring hypothesis. When scientists invoke mathematical objects, perhaps this is done without ontological seriousness, merely in order to use these objects as representational aids to facilitate the expression of various nominalistic claims that they *are* serious about. If this is right, perhaps there is no need to go through all the hard work of Field's nominalization strategy. Perhaps there is "an easy road to nominalism" that leaves science just as it is but interprets its pronouncements in a way that eliminates all commitment to mathematical objects and thus extracts the real—and purely nominalistic—content of these pronouncements. This daring hypothesis has recently been pursued with much fervor and creativity. Three brief remarks will have to suffice for our present purposes.

---

[21] See Melia (1995) and Yablo (2005).

First, once we properly understand the representation theorems at the heart of Field's approach, it is not hard to nominalize the statements from the mentioned examples and to prove appropriate representation theorems, thus subsuming these examples under Field's approach. Consider (3). We first introduce plural predicates that enable us to define a four-place plural predicate SAMERATIO(xx, yy, zz, ww), meant to express that the ratio of yy to xx is the same as that of ww to zz.[22] Then we lay down uncontroversial nominalistic assumptions governing the newly introduced predicates and assume there are infinity many objects (say spacetime points). We can now prove that there is a function $r(xx, yy)$ from two finite pluralities to a positive rational number such that

(7.1) $$\text{SAMERATIO}(xx, yy, zz, ww)$$
$$\leftrightarrow r(xx, yy) = r(zz, ww)$$

Of course, $r(xx, yy)$ is the ratio of yy to xx.[23]

Second, suppose there are cases that resist Field's approach because no nominalization is available. It is precisely in such cases that an easy road to nominalism would make a difference. Whether an easy road *would* be available in such cases, however, is debatable.[24] What would it even mean to "extract the purely nominalistic content" of some mixed physical and mathematical utterance in a case of the hypothesized sort? The examples from Malament's challenge to Field illustrate the concern (cf. §7.3).

---

[22] Recall from §6.2 that 'xx,' etc., are plural variables, allowed to have one *or more* values.

[23] Let me sketch the proof. We first introduce a plural predicate 'SUCC(xx, yy),' which intuitively states that the number of yy directly succeeds that of xx, but officially is defined nominalistically in terms of equinumerosity of pluralities (which we take as primitive). We next introduce plural predicates corresponding to addition and multiplication, as well as nominalistic versions of recursion clauses for these arithmetical relations (cf. §11.2). By a plural (and thus nominalistic) transposition of the ordinary treatment of ratios, we can now define the predicate SAMERATIO. We prove (7.1), first in the special case where yy and ww are "singleton pluralities," and then extend to the general case by means of induction.

[24] See Colyvan (2010).

What would be the purely nominalistic content of some claim about all possible states of a dynamical system or about Hilbert spaces in quantum mechanics?

Finally, suppose (if only for the sake of the argument) that the concern voiced in the previous comment can be addressed and that mathematics can thus be shown to play a merely representational and deductive role in empirical science. What would this discovery show? It would obviously undermine any version of the indispensability argument, which bases our faith in mathematics on its allegedly indispensable contributions to the empirical sciences. It is supremely important, however, to notice that the discovery would pose no immediate threat to other defenses of mathematics. The best defense of mathematics might still turn out to flow, not from its contributions to the empirical sciences, but from a more distinctively mathematical or logical source of evidence. This possibility will be explored in the chapters that follow.[25]

SELECTED FURTHER READINGS

Benacerraf (1973) is the classic presentation of the epistemological objection to object realism. Field (1982) provides a useful introduction to his nominalization program, although *Science without Numbers* (1980) remains the canonical presentation. Shapiro (1993) develops an important objection. Two "easy roads" to nominalism are developed in Melia (1995) and Yablo (2005). Colyvan (2010) criticizes these approaches; further discussion by six philosophers can be found in the October 2012 issue of *Mind*. Yablo (2014) spells out in greater detail his interesting (but complex) contribution to this issue.

---

[25] Cf. also the final paragraph of Chapter 6.

Mathematical Intuition

8.1 EVIDENCE IN MATHEMATICS

How do we come to know mathematical truths? While it would be unreasonable to impose a causal requirement on such knowledge, we still want an informative answer to the question (cf. §§1.5 and 7.1).

One answer is that mathematical knowledge is broadly empirical (cf. chap. 6). But this answer struggles to do justice to mathematics as it is actually practiced. The most sophisticated form of empiricism about mathematics is Quine's, which compares the epistemological status of mathematics to that of theoretical physics. The comparison is problematic, however. Elementary mathematics appears to enjoy a far stronger and more direct form of evidence than the often tenuous and indirect evidence enjoyed by theoretical physics.

We shall now take a closer look at some accounts of mathematical knowledge that do not assimilate it so strongly to empirical knowledge. One option came up in our discussions of Kant, Hilbert, and Brouwer, namely that some form of *mathematical intuition* provides evidence for certain mathematical truths. This idea will be our main focus in the present chapter.

There are other options too, to be discussed in later chapters. A brief overview may be useful. According to Frege and his followers, logic—or at least broadly conceptual considerations— provide a source of evidence in mathematics (cf. chap. 9). Another important issue in the epistemology of mathematics is that of extrapolation. This issue plays an important role in our set-theoretic reasoning, where some very natural ideas about collecting objects are extrapolated from their most secure home in applications to finite domains and applied to huge infinite domains (cf. chap. 10). Last, Russell and Gödel seek inspiration from the natural sciences, where a theoretical hypothesis can be supported by its capacity to systematize, explain, and predict more elementary observations. Both thinkers argue that analogous considerations can serve as evidence in mathematics, including for its higher reaches (cf. chap. 12).

When discussing the various possible sources of evidence in mathematics, two points should be kept in mind. First, the availability of one form of evidence need not preclude the availability of other forms. As we shall see, Gödel was a pluralist about mathematical evidence, defending a role for mathematical intuition, conceptual analysis, and indirect explanatory evidence. Second, evidence may come in degrees. While elementary mathematics may enjoy a particularly strong and direct form of evidence, only more attenuated forms of evidence may be available for parts of higher mathematics. For example, Parsons argues that there is such a thing as intuitive evidence—but that it does not take us beyond parts of finitary mathematics. In short, we should take seriously the possibility of a conception of mathematical evidence that is both *pluralist* and *gradualist*.

8.2 THE NOTION OF INTUITION

"Intuition" is multiply ambiguous. The term is often used simply in the sense of an immediate or pretheoretic opinion. Philosophers sometimes appeal to intuitions in this sense. Such appeals are more widespread and less controversial in linguistics, however, where we have immediate reactions concerning the grammaticality of sentences. Consider, for example, the following well-known pair (where, as usual, ungrammaticality is marked by '*'):

(4) I wish/hope that John will leave.
(5) I wish/*hope John to leave.

Another notion of intuition is that of an immediate rational insight. This notion figures prominently in the rationalists, for instance in Descartes's appeal to "the natural light of reason," which can be trusted and serve as a source of knowledge. We shall not have anything to say about these two notions of intuition.

117

We shall instead focus on a notion of intuition employed by Kant and other thinkers whom he inspired. In particular, Hilbert argues that we have intuition of what we are now calling Hilbert strokes (cf. §4.4). Since these strokes are understood as types, not tokens, they are abstract and thus cannot strictly speaking be perceived. Hilbert nevertheless contends that we have a broadly perceptual mode of access to types, provided by a form of mathematical intuition. We shall discuss some recent attempts to defend this idea.

8.3 SKEPTICISM ABOUT MATHEMATICAL INTUITION

The idea of intuitive evidence for mathematics has encountered substantial skepticism in recent philosophy.

One reason is that such evidence may seem *irrelevant*. True, mathematicians used to appeal to intuitive evidence. For example, the intermediate value theorem used to be regarded as intuitively obvious. Surely, any continuous function whose value begins below some number but ends above it must at some point have this number as its value (cf. §2.1). But Bolzano showed that a better proof is possible, replacing the appeal to intuition with a rigorous analytical argument. This development continued throughout the nineteenth century with the rigorization of analysis. Appeals to intuition gradually gave way to analytical and eventually set-theoretic arguments.

Ultimately, however, this development only pushes the epistemological question back to set theory. Might intuition play a role there? Gödel famously thought so:

> But, despite their remoteness from sense experience, we do have something like a perception also of the objects of set theory, as is seen from the fact that the axioms force themselves on us as being true. I don't see any reason why we should have less confidence in this kind of perception, i.e. in mathematical intuition, than in sense perception. (1964, pp. 483–84)

However, do we really have something like perceptual access to the vast infinite sets postulated by modern set theory? And do the set-theoretic axioms really "force themselves on us"? Many thinkers have dismissed such claims as simply incredible. So a second reason for skepticism about mathematical intuition is that this alleged capacity seems *mysterious*.

It cannot be denied that some appeals to intuition are merely "just so" stories, devoid of explanatory power. Russell's talk of "acquaintance" with universals provides an example. At one stage, Russell took mathematical propositions to be concerned with abstract universals; for example, "the proposition 'two and two are four' [...] states a relation between the universal 'two' and the universal 'four'" (1912, p. 103). This raises the question of how we acquire knowledge of such universals and their relations to one another. Russell answers that we have "knowledge by acquaintance" not only of "sensible qualities" but also of "relations of space and time, similarity, and certain abstract logical universals" (p. 109). He concludes that "all our knowledge of truths depends on our intuitive knowledge." But Russell offers no account of how this acquaintance works. Unlike sensible qualities, which he takes to be "exemplified in sense-data," the abstract logical universals are not given in sensation. How, then, do we apprehend them? Is this the work of some special mental faculty? Instead of giving proper answers, Russell simply postulates a form of acquaintance with universals with no real explanation.

In sum, to be convincing, an account of mathematical intuition must explain why such intuition is neither irrelevant nor mysterious.

8.4 Some Recent Defenses of Mathematical Intuition

Penelope Maddy has argued that we can perceive impure sets, such as the set of twelve eggs in a carton.[1] Her argument has obvious potential relevance to the epistemology of set theory.

---

[1] See Maddy (1990, chap. 2). A set is said to be *impure* when it has non-sets in its transitive closure.

And she takes great care to ensure that the form of perception she describes is scientifically respectable. The central idea is that we have an ability to represent many objects considered together as a form of unity. This unity (or set) is located where its constituents (or elements) are, thus ensuring that sets of concrete objects are themselves concrete. Maddy even advances some hypotheses about the neural mechanisms underlying our perception of such concrete sets.

As noted, Gödel expressed high hopes concerning a form of perception of sets. Maddy's account of set perception has far more limited scope. Since the objects to be considered as a unity need to be given in perception, there can on her account be no perception of pure sets, which involve no perceptible objects. It is not even clear how perception of sets of sets of concrete objects would work. (What would it be to perceive the powerset of the mentioned set of eggs?) Nor does Maddy claim that we have perceptual access to infinite sets, even when these are impure (and thus concrete).

In contrast to Maddy's naturalistic approach, Charles Parsons defends a broadly Kantian conception of mathematical intuition. Such intuition, he argues, presents us with types of perceptible tokens in a way that is "strongly analogous" to how ordinary concrete objects are given in perception; he therefore calls such intuition *quasi-perceptual* (Parsons, 1980, p. 162). If correct, Parsons' claim has great significance for the epistemology of finitary mathematics. As Hilbert realized, finitary mathematics can be understood as concerned with syntactic types, such as Hilbert strokes, and their basic syntactic properties (cf. §4.4). An intuitive mode of access to such objects would therefore provide a particularly strong and direct form of evidence for this part of elementary mathematics—in stark contrast to the more attenuated forms of evidence available for set theory, to which finitary mathematics could otherwise be reduced.

Notice that Parsons goes beyond Maddy by defending a form of intuition of objects that are not concrete. As we shall see, however, it is important for his account that the syntactic types that we intuit be at least "quasi-concrete"—in the sense that they have canonical instantiations in spacetime.

Why believe that we have intuition of types, not just perception of corresponding tokens? Parsons's argument seeks to establish that types play an important role in our sensory experience. Our identification of a type is often firmer and more explicit than the identification of the corresponding token. Consider, for example, episodes of perceiving the word 'dog'. Very often, we do not notice the accent with which the token was pronounced, the font with which it was written, or any other property that would distinguish one token of the word from another. What registers most firmly and explicitly in our minds is only a phonetic or orthographic type. In cases such as these, our experience seems to be directed at a type, not a token.

This is a plausible description of how things *seem* to the subject. One might nevertheless worry that this form of intuition would be mysterious. Would it not require an entirely new mode of access to reality in addition to our familiar and fairly well understood senses? Parsons's response tries to "dispel the widespread impression that mathematical intuition is a 'special' faculty, which perhaps comes into play only in doing pure mathematics" (1980, pp. 154–55). Borrowing an idea from Husserl, he proposes that intuition is always "founded" on ordinary sensations or imaginings. That is, we perceive or imagine something particular, which is an instance of the more abstract object of our intuition. By anchoring intuition in this way to capacities that we indisputably have, there is no need for any mysterious "special faculty" of intuition.

One difference between mathematical intuition and ordinary perception is thus that the former is "founded." This difference has an important consequence concerning the possible objects of mathematical intuition. For the required "founding" to be possible, the objects of intuition must be quasi-concrete. It is this that allows the intuition to be "founded" on perceptions or imaginings of corresponding tokens. So by requiring that mathematical intuition be "founded," Parsons also limits its scope. We cannot have intuition of abstract objects that do not have tokens in space and time, such as numbers or pure sets, which are not quasi-concrete but purely abstract.

Another difference is that mathematical intuition relies more heavily on conceptualization than ordinary perception does. As Parsons observes, "*[w]hat is intuited* depends on the concept brought to the situation by the subject" (1980, p. 162). For example, one cannot intuit linguistic types without having at least an implicit grasp of the difference between types and tokens and of how the relevant types are individuated. Mathematical intuition therefore requires more sophistication and training than ordinary perception. This observation goes some way toward explaining why our capacity for mathematical intuition is so much less obvious to us than our capacity for ordinary perception. Without the training needed to acquire the relevant concepts, we cannot enjoy the benefits of such intuition.

Where Parsons draws on Kant and Hilbert, Dagfinn Føllesdal looks to Husserl in an attempt to develop a phenomenological conception of mathematical intuition.[2] Føllesdal first asks us to consider Jastrow's famous duck-rabbit drawing, which can equally well be seen as a duck and a rabbit, depending on what we focus on. The drawing thus shows that "we can experience a multitude of different objects when we are in a given sensory situation: a duck, a rabbit, but also an ear of a rabbit, an eye, or even the color or the front side of an object" (Føllesdal, 1995, p. 429). Our perception is an active process, where we choose what to focus on and are actively involved in interpreting the sensory information that we receive. This active character of our perception suggests that we can equally well choose to focus on an object's more abstract features, such as its shape. When we stand in front of a tree with a nice triangular shape, for example, we can choose to concentrate on its shape. Doing so results in an intuition of triangularity.

Føllesdal's account of intuition has much in common with Parsons's. Both emphasize that such intuition is "founded" on acts of perceiving or imagining; both accounts are closely connected with the distinction between types and tokens; and both acknowledge the reliance of intuition on concepts and concep-

---

[2] He also draws on Gödel, whose view on mathematical intuition he shows to be influenced by Husserl's.

tualization. But Føllesdal places more emphasis than Parsons does on our own contribution to the structuring of what we perceive or intuit. This difference emerges most clearly when we consider concepts that are more abstract than that of triangularity. Consider the concept of *topological genus*, which can be thought of as the number of holes that an object has. An example will help to convey the idea. A cup with a handle is superficially far more similar to a glass than to a donut: both the cup and the glass are shaped so as to serve as drinking vessels. But let us use our imagination to explore how an object can be transformed into another in a continuous manner, that is, by compressing, stretching, and twisting, but with no tearing. Since the cup shares with the donut the property of having a single hole, we can imagine the former being continuously transformed into the latter.[3] By contrast, the glass cannot be continuously transformed to the cup, since a hole would at some point have to be torn in the glass to make a handle. With some training, we can in this way become perceptually aware of a highly abstract property that the cup shares with the donut but not the glass, despite the cup's far greater superficial similarity to the glass than to the donut.

Let us take stock. We have found that a strong case can be made for the existence of a broadly perceptual source of evidence about mathematical objects that are either concrete (as Maddy thinks is the case with impure sets) or quasi-concrete (as Parsons's types). If defensible, this source of evidence will play an important role in the epistemology of elementary mathematics. However, we have found reasons to doubt that this broadly perceptual evidence will provide much support for what Hilbert calls "infinitary mathematics." In order to find evidence for higher mathematics, it appears we must look elsewhere. Two options were mentioned in §8.1. We may analyze our mathematical concepts, such as that of set. Or we may seek indirect evidence for mathematical axioms via their ability to systematize and explain more elementary observations. These options will be explored in Chapters 10 and 12, respectively.

---

[3] Should your imagination fail you, a web search on "animation, cup, donut" will enable you to *perceive* the transformation instead.

8.5 MATHEMATICAL INTUITION AND CRITERIA
   OF IDENTITY

The accounts of mathematical intuition proposed by Parsons and Føllesdal seem quite plausible from the point of view of the perceiving subject. But there is a residual worry. Do the accounts succeed in explaining our access to abstract objects in a way that meets the integration challenge (cf. §1.5)?

Let us grant that we can perceive a ball to be round and some marbles to be five in number. This means that properties and simple structural features can be present in our perception. But it does not establish that we can perceive abstract objects. In the mentioned examples, the proper objects of perception are the ball and the marbles, respectively. Roundness and "fiveness" are merely *predicated of* these objects and are not themselves the proper objects of perception. The idea of mathematical intuition promises the existence of quasi-perceptual acts in which such properties and structural features figure as the proper objects of perception and thus also as the subjects of other predications. An example would be an intuition that roundness is distinct from being cubical.

The required shift from predicables to proper objects of predication is an instance of what philosophers call *reification*; that is, coming to regard an item as *an object*. Now, wherever there are objects, it must make sense to ask questions about identity and distinctness: is *this* object identical with *that*? (As Quine famously put it, there can be "no entity without identity.") This marks an important shift. Questions about the identity and distinctness of properties were not obligatory prior to their reification, when they were merely put to predicative use. For example, we can perceive that the ball is round without taking a stand on whether roundness is identical with some other property which we perceive another object to have (say being a good but not perfect approximation of a sphere). This neutrality is no longer possible when roundness is recognized as an object.

What does a question about the identity and distinctness of properties turn on? How does the world have to be in order to make the relevant identity statement true? *Criteria of identity* are supposed to systematize our answers to such questions. Cardinality properties, such as "fourness," provide a nice example. Is the cardinality of *these* things identical with the cardinality of *those*? It is plausible to take the answer to be 'yes' just in case the former things can be one-to-one correlated with the latter. That is, Hume's Principle is plausibly taken to provide a criterion of identity for cardinality properties (cf. §2.4).

We are now in a position to connect the above discussion with another topic that figures prominently in this book, namely abstraction (cf. §2.4). We began by noting that we perceive and imagine objects to have various properties and stand in various relations. We then asked what is required for such properties to be regarded as objects in their own right. It is widely believed that we need a criterion of identity to specify when two property instantiations count as instantiations of the same property. But to specify when two objects count as instantiating the same property is nothing other than to provide an equivalence relation on objects; that is, a relation that is reflexive, symmetric, and transitive. The discussion in this chapter shows that our handle on this equivalence relation can be implicit and perceptual, rather than explicit and conceptual. In cases where we rely on perception to determine whether two objects stand in the equivalence relation, it seems reasonable to talk about a quasi-perceptual mode of access to the ensuing properties. But in other cases, perception plays little or no role. In the next chapter, we shall therefore undertake a general examination of how equivalence relations—serving as criteria of identity—can underlie our apprehension of abstract objects.

Selected Further Reading

The approaches to mathematical intuition discussed in this chapter are clearly presented in Maddy (1990, chap. 2); Gödel (1964); Parsons (1980); and Føllesdal (1995). Parsons (2008, chap. 5) is a more complete exposition of Parsons' approach. Tieszen (1989) provides a fuller development of a Husserlian approach.

Abstraction Reconsidered

9.1 A Simple Example of Abstraction

Abstraction appears to play an important role in our mathematical thought. Whenever we have an equivalence relation, we appear to be able to talk about the abstract feature that two entities share just in case they stand in the equivalence relation. Consider, for example, the equivalence relation that obtains between two pluralities of things just in case the pluralities are equinumerous (that is, can be correlated one-to-one). We appear to be able to talk about the abstract feature that any two equinumerous pluralities share, namely their cardinal number. Unfortunately, there are dark clouds on the horizon. Frege's pioneering work on abstraction ended in paradox (cf. §2.7).

Before preparing for the storm, some reconnaissance will be useful. We shall therefore begin by taking a closer look at a family of particularly clear and promising cases of abstraction. One of Frege's favorite examples concerns directions. Consider a domain of lines on which the equivalence relation of parallelism is defined. This equivalence figures in the criterion of identity for directions:

(Dir) $$d(l_1) = d(l_2) \leftrightarrow l_1 \parallel l_2$$

where '$d(l)$' stands for the direction of the line l. I shall say that two lines *specify* the same direction just in case they are parallel. On the plausible assumption that lines and facts about parallelism are epistemically accessible to us, this criterion enables us to talk about directions being identical or distinct. What about other properties and relations? Orthogonality provides an example. Two directions are regarded as orthogonal just in case they are specified by two orthogonal lines. Using $\perp$ and $\perp^*$ as orthogonality predicates for lines and directions, respectively, we formalize this as follows:

(9.1) $$d(l_1) \perp^* d(l_2) \leftrightarrow l_1 \perp l_2$$

The orthogonality of two directions is, as it were, "inherited" from the orthogonality of any two lines in terms of which these directions are specified.

Of course, this "inheritance" of properties from lines to the directions that the lines specify presupposes that it doesn't matter which line we choose in order to specify some given direction. Thankfully, the presupposition is met. Assume that the direction specified by l_1 is also specified by l_1', because $l_1 \parallel l_1'$. Then, if one of the two specifications is orthogonal to l_2, so is the other. For the purposes of assessing orthogonality, any line is just as good as its parallels. Moreover, the example generalizes. For any predicate P on lines that doesn't distinguish between parallel lines, we can introduce an associated predicate P^* on directions by letting P^* hold of the directions of some given lines just in case P holds of the lines themselves.

Where does this leave us? Assume that our discourse about directions is governed by the mentioned rules and their generalizations to other predicates. As Frege observes, it is then precisely *as if* we are talking about directions as objects. The rules allow us to talk about directions—just as more familiar objects— as presented in different ways, as identified and distinguished, and as objects of various predications. At the very least, this shows how it can be permissible to talk *as if* there are abstract mathematical objects—in this case, directions.

Do directions really exist, however? One may be inclined to answer "no." All that we have done, it seems, is to show how one can legitimately and truthfully *speak like* a platonist without actually being one. This impression can be sharpened. Using the mentioned clauses and generalizations, it can be shown that any statement seemingly about directions can be translated into a statement concerned only with lines. This suggests that all we have achieved is to define a *manner of speaking* that sounds platonist but in fact is nominalistically acceptable. All we have secured is the appearance of platonism, not the reality of it.

Frege and many of his followers see things differently. If we can legitimately and truthfully speak as if there are abstract objects such as directions, what more might be required for such objects to exist? There is no higher standard for the existence of objects of some sort than that which governs our discourse about such objects. To require more would be to hold this discourse to an unreasonable standard, imposed on it from the outside. Fregeans defend an alternative conception of what has been achieved. We start with a domain of entities standing in certain relations. Claims about these entities are then "reconceptualized" as claims about certain abstract features that two such entities have in common just in case they stand in some appropriate equivalence relation.[1] For instance, a claim about parallelism of lines is "reconceptualized" as a claim about identity of directions. What is required for the existence of the directions is thus nothing over and above the existence of lines standing in appropriate relations of parallelism. There is no "metaphysical distance" between the former fact and the latter. Likewise, the orthogonality of two directions is nothing over and above the orthogonality of the two lines in terms of which the directions are specified.

If defensible, this alternative Fregean conception will explain our "access" to abstract objects such as directions and thus allow us to meet the integration challenge (cf. §1.5).[2] Directions are specified by means of lines, which are assumed to be unproblematic. And since all the properties and relations of directions are "inherited" from corresponding properties and relations of lines, these don't pose any additional epistemological problem. There is no "distance" between the two sets of facts. One is obtained by a "reconceptualization" of the other.

---

[1] Where Frege talked about "recarving of meaning" (cf. §2.4), I shall now use the term "reconceptualization," which I find more apt. This also signals that our primary aim is now to develop some Fregean ideas, not exegesis.

[2] This conception will also fill the gap in accounts of mathematical intuition that was identified in §8.5.

9.2 THE THREAT OF PARADOX

We turn now to the paradoxes that threaten the use of abstraction. The problem arose, we recall, because Frege sought to reduce all forms of abstraction to a single form. When two concepts are coextensive, let us say that they have the same extension. This yields Frege's infamous "Basic Law V," which in contemporary notation can be written as

(BLV) $\qquad \{x \mid Fx\} = \{x \mid Gx\} \leftrightarrow \forall x (Fx \leftrightarrow Gx)$

As Russell discovered, this "law" gives rise to the paradox now bearing his name (cf. §2.7). To recall how, it is useful to think of Frege's extensions as classes. The "law" then allows us to consider the Russell class r whose members are each and every object that is not a member of itself. We now ask whether r is a member of itself. It is easy to derive the contradiction that r is a member of itself just in case it is not.

What to do? Given its importance to our mathematical thought, it would be an overreaction simply to give up on abstraction, as Frege appears to have done around 1906. In the next three sections, we shall discuss three more ambitious responses. All three recognize an indisputable lesson from Russell's paradox, namely that it can be dangerous to abstract on concepts in a way that yields objects. According to the first approach—advocated by Russell and Alfred North Whitehead—we need to find an altogether different way of accounting for abstraction, which avoids the perilous transition from concepts to objects. A second approach—advocated by the neo-Fregeans Bob Hale and Crispin Wright—accepts only selected transitions from concepts to objects. Although Basic Law V is impermissible, other forms of abstraction are unproblematic. Clearly, an important challenge for this approach will be to clarify the distinction between permissible and impermissible forms of abstraction. A third and final approach bypasses this challenge by accepting all forms of abstraction, while avoiding paradox by allowing the objects obtained by abstraction to lie outside of the domain on which we abstract—much like the directions, in our motivating example, lie outside of the domain of lines.

9.3 THE SIMPLE THEORY OF TYPES

After discovering the paradox, Russell, along with his collaborator Whitehead, developed an alternative to Frege's inconsistent theory of classes or extensions, resulting in the three-volume behemoth *Principia Mathematica*, published in 1910–13. Their proposed theory is notoriously complex. We shall consider here only a substantially simplified version known as *the simple theory of types* (STT), brought to prominence by the precocious F. P. Ramsey (1903–30) in his (1931).

STT distinguishes sharply between individuals, classes of individuals, classes of classes of individuals, etc. As Russell suggests, we may think of these as respectively individuals, families, clans of families, etc. In fact, the distinction between individuals and classes of the various levels is so sharp that each level has its own set of variables and constants, each with a superscript that indicates its level or type (as it is also called). For example, variables for individuals and for "clans" are of type 0 and 2, respectively. Furthermore, a membership claim $s \in t$ is considered well-formed only if the type of t is exactly one higher than that of s. Thus, while it is permissible to say that an individual is a member of a family, or a family of a clan, it is impermissible to say that an individual is a member of a clan or of another individual. It follows that the condition meant to define the Russell set, namely '$x \notin x$,' is ill-formed. Russell's paradox is thus blocked by a grammatical prohibition. The same goes for all the other set-theoretic paradoxes. So at least in this regard, STT is successful.

Another virtue of STT is that it allows us to handle abstraction in the way that came to dominate among mathematicians toward the end of the nineteenth century, namely by means of equivalence classes. Consider, for example, the equivalence relation of two families' being equinumerous. A cardinal number is supposed to be some form of property that two families share just in case they are equivalent in this way. So why not simply identify the cardinal number of a family with the clan consisting of all families equinumerous with this given family? This is mathematically elegant, and it ensures that two families have the same cardinal number just in case they are equinumerous.

As a logist account of mathematics, however, STT is now generally regarded as a failure. One problem is that the strict division into types is cumbersome and needlessly restrictive.[3] A related problem is that its syntactic restrictions on membership claims block not only the paradoxes but also Frege's celebrated *bootstrapping argument*, which shows that there are infinitely many numbers. The argument crucially turns on the fact that numbers too can be counted. Thus, if we have established the existence of numbers 0 through n, we have at least $n + 1$ objects available to count, which enables us to establish the existence of the number $n + 1$ as well. This argument is no longer available when numbers are construed in Russell and Whitehead's way as clans of equinumerous families. As a result, we need a separate *axiom of infinity*, which states that there are infinitely many individuals. But it is doubtful that this axiom can be regarded as a purely logical principle. It certainly does not qualify as logical on our contemporary conception of logic, which requires a logical truth to be true in all models. Russell and Whitehead thus found themselves in need of assumptions with substantive mathematical content, which is not a happy situation for aspiring logicists.

The real value of STT, in my opinion, is not the intended one of providing a purely logical foundation for mathematics but rather derives from STT's role as a simple and weak set theory and as an important steppingstone toward stronger and more interesting set theories (cf. §10.3).

9.4 NEO-FREGEAN ABSTRACTION

To explain the neo-Fregean view, a brief review of Frege's account of arithmetic may be useful (cf. §§2.4 and 2.6). The account proceeds in two steps.

The first step consists of an account of the applications and identity conditions of numbers. Frege argues that counting involves the ascription of numbers to concepts. For instance,

---

[3] An argument of Gödel's to that effect will be discussed in §10.3.

when we say that there are eight planets, we ascribe the number eight to the concept "...is a planet." Frege's claim is that the number-of operator '#' applies to any concept expression F to form the expression '$\#x\,Fx$,' meaning "the number of Fs." Next Frege argues that the number of Fs is identical to the number of Gs if and only if the Fs and the Gs can be put in a one-to-one correspondence. This is known as *Hume's Principle* and is formalized as

(HP) $\qquad\qquad \#x\,Fx = \#x\,Gx \leftrightarrow F \approx G$

where $F \approx G$ is a formalization in pure second-order logic of the claim that the Fs and the Gs are equinumerous.

The second step seeks to provide an explicit definition of terms of the form '$\#x\,Fx$.' In order to do so, Frege uses a theory consisting of second-order logic and Basic Law V. He defines $\#x\,Fx$ as the extension of the concept "is an extension of some concept equinumerous with F."[4] It is straightforward to verify that this definition satisfies (HP).

How should we respond to the well-known fact that Frege's approach is inconsistent? A simple but radical answer is proposed by Wright (1983). Why not simply abandon the second step of Frege's approach—which introduces the inconsistent theory of extensions—and make do with the first step? This proposal has sparked a neo-Fregean approach to the philosophy of mathematics, developed in large part by Wright in collaboration with Bob Hale.

The neo-Fregean proposal relies on two fairly recent technical discoveries. The first discovery is that (HP), unlike Basic Law V, is consistent. More precisely, let *Frege Arithmetic* be the second-order theory with (HP) as its sole nonlogical axiom. Frege Arithmetic can then be shown to be consistent if and only if second-order Dedekind-Peano Arithmetic is (cf. §2.6).[5] The second discovery is that Frege Arithmetic and some very

---

[4] That is, $\#x\,Fx$ is defined as $\{x \mid \exists G(x = \{y \mid Gy\} \land F \approx G)\}$.

[5] *Proof sketch*: Let the domain D consist of the natural numbers. If a concept F applies to n objects, let '$\#x\,Fx$' refer to $n + 1$. If F applies to infinitely many objects, let '$\#x\,Fx$' refer to 0.

natural definitions suffice to derive all the axioms of second-order Dedekind-Peano Arithmetic. This result is known as *Frege's theorem*. So at least from a technical point of view, the neo-Fregean approach is a success: it is consistent and strong enough to prove all of ordinary arithmetic.

What about the philosophical merits of the approach? Let us begin by asking why Frege insisted on the second of the two steps described above. One reason may have been the greater generality of the resulting approach. Extensions enable us to imitate not only (HP) but any other form of abstraction on concepts as well. But Frege's official reason is different. While (HP) gives us a handle on all identity statements of the form '$\#x\,Fx = \#x\,Gx$,' the principle is silent on mixed identity statements such as '$\#x\,Fx = $ Julius Caesar.' Perhaps we know that all such statements are false; but if so, this is no thanks to (HP). This is known as *the Caesar problem*. Frege thought we have a firmer grasp on extensions, which enables us to distinguish extensions from Caesar and all other concrete objects.[6]

If we want to abandon the second step, we need an alternative solution to the Caesar problem. Hale and Wright propose a solution based on a simple but powerful idea. When we learn that a criterion of identity applies to an object, we learn something about the object, namely that it has a certain property. So when a criterion of identity applies to one object but not to another, the former object has a property that the latter lacks. It follows by Leibniz's Law that the objects are distinct. While the details obviously need to be spelled out, this is a promising beginning.[7]

A second question concerns the philosophical status of Hume's Principle. As a logicist, Frege took Basic Law V to be a logical truth. But Russell's paradox proved him wrong. Might we retreat to the claim that (HP), at least, is logically true? The problem is that (HP) doesn't much look like a logical truth.

---

[6] At least this is the impression we get from Frege (1953, §68). However, a form of the Caesar problem for extensions is recognized in Frege (2013, §10), thus limiting the advantage of the second step to its greater generality.

[7] See Hale and Wright (2001b) for further details.

Our contemporary conception requires a logical truth to be true in all models. But (HP) is true only in infinite models. The neo-Fregeans therefore settle for the weaker claim that (HP) is an *a priori* principle "whose role is to explain, if not exactly to define, the general notion of identity of cardinal number" (Hale and Wright, 2001a, p. 279). Their attempts to defend this weaker claim are based on two different but related ideas.

One idea has already been mentioned, namely that the identity statement on the left-hand side of (HP) is merely a "reconceptualization" of the content of the equinumerosity statement on its right-hand side (cf. §§2.4 and 9.1). As Frege puts it, the two statements are just different ways of "carving up" one and the same content (1953, §64). Many philosophers agree that this is a tantalizing idea. But the idea has proved difficult to substantiate, and its prospects are still being debated.[8]

Another idea is that (HP) serves as an implicit definition of the number-of operator #.[9] More precisely, we have an *a priori* entitlement to lay down (HP) as an implicit definition; and if the definition succeeds, this will give us *a priori* knowledge of numbers and their properties. It is clear, however, that not all implicit definitions offer such substantial epistemological benefits. Consider the implicit definition of 'Jack the Ripper' as the person, whoever he might be, who committed certain gruesome murders. For the definition to succeed, there must be a single individual who committed the murders. This is a substantial presupposition. So at best, the definition may provide *a priori* knowledge of the conditional that *if* the presupposition is satisfied, *then* Jack committed the murders. Why should (HP), understood as an implicit definition, do any better? Shouldn't any *a priori* knowledge flowing from this definition be conditional on the equally substantial presupposition that there is in fact a function that maps concepts to numbers in accordance with (HP)? But this highly conditional knowledge falls far short of outright knowledge of arithmetic.

---

[8] See Field (1984) for a classic criticism and Rayo (2016) for a recent defense.

[9] See Hale and Wright (2000).

The neo-Fregeans respond that (HP) is free of presuppositions because it serves as a criterion of identity for numbers. Here is the idea. Since (HP) serves as a criterion of identity, the statements that flank its biconditional are intimately related. The equinumerosity $F \approx G$ is "conceptually sufficient" for the truth of the identity $\#x\, Fx = \#x\, Gx$. There is "no gap" between the two statements that some metaphysical assumption is needed to "plug" (Hale and Wright, 2009, p. 193). Notice, however, that this response leads us back to the first idea concerning "reconceptualization." So when the second idea is developed in this way, it is not independent of the first.

The third and final question that we shall consider is which abstraction principles are permissible. The answer will have to balance two conflicting pressures. On the one hand, the rot represented by Basic Law V and a wide variety of other impermissible abstraction principles will have to be excised in a way that is both definitive and well-motivated. On the other hand, we want to leave behind not only Hume's Principle but ideally also other abstraction principles that can serve as a foundation for analysis and set theory. This requires a careful balancing act. Thanks to a flurry of recent activity, substantial progress has been made, especially on the technical side. Abstractionist approaches that take us well beyond arithmetic have been developed.[10] We also have a far better understanding of the different ways of excising the rot and of their respective strengths and weaknesses.[11] But some hard philosophical work still remains before we have an adequate demarcation of the permissible abstraction principles from the impermissible. The task of providing such a demarcation is not solely a technical one but should also be integrated with our philosophical account of how abstraction works. Our philosophical account should motivate, or at least inform, the answer to the demarcation problem.

---

[10] See, e.g., Hale (2000) on real analysis and Shapiro (2003) on set theory.

[11] See Burgess (2005) and, for an overview, Linnebo (2009).

9.5 Dynamic Abstraction

A recently developed approach to abstraction gives a strikingly different answer to the demarcation problem.[12] Rather than try to distinguish the good abstraction principles from the bad, this approach accepts them all—albeit in a special manner. Let me explain. Consider a domain of entities on which we want to abstract. These entities can be either objects or Fregean concepts based on some domain of objects. On the neo-Fregean approach, the abstract objects obtained by some form of abstraction are always assumed to belong to the domain of objects with which we began. Neo-Fregean abstraction is in this sense "static." It always takes place relative to some fixed domain of objects.

This static conception is not obligatory. There is, for example, no need to assume that the directions obtained by abstraction on lines under the equivalence of parallelism belong to the domain with which we started (cf. §9.1). There is also a "dynamic" conception on which abstraction may result in "new" objects that lie outside of the "old" domain with which we began. In fact, this dynamic conception of abstraction is arguably *more* congenial to our paradigm examples of abstraction. The distinctive feature of these examples is that every question about the objects obtained by abstraction reduces to—and thus is "reconceptualized" as—a question about the entities on which we abstract. For example, the question about the orthogonality of two directions reduces to a question about the orthogonality of any two lines in terms of which the directions are specified. But we must be careful. How do we know that the reductions always lead to simpler questions, which will thus eventually receive answers? The dynamic conception offers a simple but powerful assurance. Every question about the "new" objects reduces to a question about the "old" objects, which we may thus assume to have been answered.

Basic Law V provides a good illustration. Suppose we ask whether the extension of one concept is identical with that of another. The strategy is to reduce this to the question of whether the two concepts are coextensive. But is the "reduced" question

[12] See Linnebo (forthcoming) and Studd (2016).

really any simpler than the initial one? On the static conception, it need not be. The question about the coextensiveness of concepts may on this conception lead us back to the initial question about the identity of extensions. For among the objects with which the "reduced" question is concerned are the very extensions whose identity or distinctness we set out to determine.[13] By contrast, no such circularity would arise if the extensions generated by Basic Law V lay outside of the domain with which we began. An identity question concerning these "new" extensions would then always reduce to a coextensionality question relative to the "old" domain, which we may thus assume to have been answered.

How do the static and dynamic approaches to abstraction compare with regard to the development of mathematics? The dynamic approach is more permissive in some respects. As we have seen, this approach permits abstraction on any equivalence relation, including coextensionality, as in Basic Law V. The dynamic approach also allows *iterated application* of each form of abstraction. Consider again Basic Law V. One application of the law takes us from some initial domain to a larger one. Since this larger domain gives rise to more Fregean concepts than the initial one, a second application of the law will give rise to even more objects. We can continue in this way indefinitely. At limit stages, we take the union of all the objects generated thus far. Since each round yields something new—as Russell's paradox would otherwise reemerge—the process will never terminate.

In other respects, the dynamic approach appears more restrictive than the static one. Hume's Principle provides an example. In its familiar static form, this principle supports Frege's famous bootstrapping argument (cf. §9.3). By counting the numbers from 0 to n, we establish the existence of the number $n + 1$. We thus prove by mathematical induction that our fixed domain contains all the natural numbers. Consider now the dynamic version of the principle. Given the numbers from 0 to n, we can

---

[13] For example, suppose that each of two concepts applies solely to its own extension. Then the coextensionality of the concepts will depend on the identity of their extensions.

again establish the existence of the number $n + 1$—but now this number may lie outside of the domain with which we began. This means that no finite number of applications of Hume's Principle will establish the existence of an infinite domain. The solution is to iterate applications of the principle. This leads to arbitrarily large finite domains and eventually also to an infinite domain.

In the next chapter, we shall discuss the iterative conception of sets, which has much in common with the dynamic approach to abstraction. On this conception, sets are "formed" in stages from objects that are "available" at the preceding stages. Some interesting questions that remain concerning dynamic abstraction have analogues concerning the iterative conception. How should we describe and theorize about iterated abstraction? Since this iterative process never terminates, the objects to which it gives rise do not form a complete totality. How, then, should we understand this incomplete totality and quantification over it?

SELECTED FURTHER READING

The further readings listed for Chapter 2 are relevant here too, especially those on the neo-Fregean approach. The neo-Fregean conception of mathematical objects is further developed in Hale and Wright (2009). The collection of essays by Hale and Wright (2001a) spells out other aspects of the approach and answers objections; students may find the introduction and Wright (1997) particularly useful. MacBride (2003) provides a useful survey of the philosophical debate concerning neo-Fregeanism, while Burgess (2005) provides a logical and mathematical analysis of various forms of abstraction. Studd (2016) is a good introduction to the dynamic approach to abstraction.

The Iterative Conception of Sets

10.1 How Sets Are "Formed"

According to the *iterative conception*, sets are formed in stages. We begin at stage 0 with no objects whatsoever.[1] At stage 1 we form all the sets of objects that are available at stage 0. Since these newly formed sets are also objects, we now have more objects available than we did at stage 0. This suggests a repetition of the operation of forming all sets of objects that are available, which will take us to an even more populous stage 2. Now we continue in this way. If at some stage α we have formed a collection V_α of sets, then at the next stage $\alpha + 1$ we form all the subsets of V_α. The domain at stage $\alpha + 1$ is therefore the union of the previous domain and its powerset; that is, $V_{\alpha+1} = V_\alpha \cup \wp(V_\alpha)$.[2] This leaves only the question of what to do at limit stages, such as stage ω, which are not immediately preceded by any other stage. The natural answer is that we simply pool all the collections of objects already formed. That is, for a limit ordinal λ, we define $V_\lambda = \bigcup_{\gamma < \lambda} V_\gamma$.[3]

The iterative conception of sets brings valuable mathematical and philosophical insights. We obtain a fairly clear picture of what set theory is about, namely the *cumulative hierarchy* that results from this stagewise formation of sets. And this picture motivates many of the axioms of today's most standard set theory, ZFC, which is adequate for almost all ordinary mathematics. (The theory will be explained shortly.) A simple example is the Pairing axiom, which says that for any two objects a and b,

---

[1] The account is easily modified to accommodate non-sets (or *Urelemente*), which may be available either at stage 0 or themselves be formed in stages.

[2] *Exercise*: Prove that this is equivalent to defining $V_{\alpha+1} = \wp(V_\alpha)$.

[3] A limit ordinal is an ordinal, such as ω, with no immediate predecessor.

there is a pair set $\{a, b\}$. Let α and β be the least stages at which a and b become available, respectively. We may assume without loss of generality that $\alpha \leq \beta$. Thus, at stage β, both objects are available. Consequently, their pair set is formed at the next stage, $\beta + 1$. Many of the other axioms are motivated in an analogous way. The iterative conception thus shows that "there is a thought behind" our standard set theory ZFC: the theory isn't just a ragbag of axioms but describes a natural structure.[4] Moreover, since we have a reasonably clear grasp of this structure, we have some evidence for the consistency of the theory.

While the iterative conception promises these important rewards, it also raises some hard questions.

1. Precisely which axioms of ZFC does the conception motivate, and how does it do so?

It would be good to give the conception a formulation that is precise enough to answer these questions. Next, the usual explanation of the iterative conception makes extensive use of temporal and constructive language. The sets are said to be "formed" in "stages" that are ordered as "before" or "after" one another. Unless one is a constructivist of some sort, this language cannot be understood literally.

2. How, then, should the temporal and constructive language be understood?

Finally:

3. Does the iterative conception single out a unique intended interpretation of the language for set theory?

One of the chief architects of the conception, Kurt Gödel, appears to have thought so. But like many of Gödel's philosophical beliefs, this thought too is controversial. The first two questions will be addressed below, and the third, in Chapter 12.

---

[4] See Boolos (1971, p. 219). However, Boolos denies that this structure is fully described by the iterative conception. Some more ambitious views on the iterative conception will be considered below.

10.2 ZERMELO-FRAENKEL SET THEORY

The discovery around the turn of the twentieth century that naive set theory is inconsistent shook the mathematical community (cf. §4.2). In the decades that followed, confidence was gradually restored. This period saw the development of two different schools that aimed to salvage some insights from the shipwreck of naive set theory. One is the type-theoretic tradition initiated by Russell and Whitehead (cf. §9.3). Their Simple Theory of Types relies on a typed language, where we have separate variables for individuals, classes of individuals, classes of classes of individuals, and so on. Thus, as Russell observes, we have one type of variables that range over individuals, another type that range over families, a third over clans, and so on. For each type n, the theory has an unrestricted comprehension principle which says that any formula $\varphi(x^n)$ defines a class $\{x^n \mid \varphi(x^n)\}$ of type $n + 1$.

Simultaneously, but independently, another school emerged among mathematicians. This school pursued a strategy that initially seems diametrically opposite to the type theorists'. They held on to an untyped language but accepted that set comprehension needs to be restricted: not every formula defines a set. An important early contribution by Zermelo (1908) formulates principles of set existence—such as Pairing, Powerset, Union— that are deemed to be *both* safe from paradox *and* needed in order to do ordinary mathematics. With assistance from Abraham Fraenkel and Thoralf Skolem, this eventually gave rise to modern set theory and its standard axiomatization ZFC.

It is important to have some familiarity with this theory. It uses a single nonlogical predicate, '$\in$' for membership. All other set-theoretic notions are defined in terms of this predicate. The axioms are as follows.

Extensionality: Coextensive sets are identical. That is, $\forall u(u \in x \leftrightarrow u \in y) \rightarrow x = y$.

Empty set: There is an empty set. That is, $\exists x \forall u(u \notin x)$.

141

Pairing: Every two objects have a pair set. That is, $\forall x \forall y \exists z \forall u(u \in z \leftrightarrow u = x \vee u = y)$.

Union: For every set z, there is a set y whose elements are precisely those objects that are an element of some element of z. That is, $\forall z \exists y \forall x[x \in y \leftrightarrow \exists w(x \in w \wedge w \in z)]$.

Power: Every set has a powerset. That is, $\forall z \exists y \forall x(x \in y \leftrightarrow x \subseteq z)$.

Infinity: There is an infinite set, that is, a set with $\varnothing$ as an element and such that, whenever x is an element, so too is $x \cup \{x\}$. That is, $\exists y[\varnothing \in y \wedge \forall x(x \in y \rightarrow x \cup \{x\} \in y)]$.

Separation: For any set z and any condition φ, there is a set of precisely those elements of z that satisfy φ. That is, $\forall z \exists y \forall x(x \in y \leftrightarrow x \in z \wedge \varphi)$.[5]

Foundation: Every nonempty set x has an element that is disjoint from x. That is, $\forall x(x \neq \varnothing \rightarrow \exists y(y \in x \wedge x \cap y = \varnothing))$.

The set theory based on these axioms is known as *Zermelo set theory*, Z. Fraenkel and Skolem argued successfully that we should add an axiom scheme that we shall now describe, which results in a theory known as *Zermelo-Fraenkel set theory* (without Choice), or ZF.

Replacement: For every set z and functional condition ψ, there is a set of precisely those objects that are borne ψ by some element of z.[6] That is

$$\text{Func}(\psi) \rightarrow \forall z \exists y \forall x[x \in y \leftrightarrow \exists w(w \in z \wedge \psi(w, x))]$$

This is based on a simple and intuitive idea. Consider any set. For each of its elements, choose either to keep this element or to

---

[5] This is an axiom *scheme*, which yields an axiom for each φ. The same goes for Replacement, stated below.

[6] ψ is *functional* just in case for every x there is a unique y such that $\psi(x, y)$.

replace it with some other object. Then the resulting collection is also a set.

ZFC is the result of adding to ZF the axiom of Choice, once highly controversial but now widely accepted.

Choice: Every set z of nonempty disjoint sets has a choice set, that is, a set containing precisely one element of each element of z.

Intuitively, given any bag containing some nonempty bags, we can produce another bag containing precisely one member of each of the bags contained in the initial bag.

As mentioned, ZFC was formulated as a result of a careful balancing act between the opposing pressures of guarding against paradox and satisfying the needs of ordinary mathematics. So it is natural to worry that the resulting set theory is *ad hoc* and devoid of intrinsic appeal. The first person who clearly saw that this isn't so was Gödel, who argued that the newly developed set theory can be motivated by a clear and natural conception, namely what we now call *the iterative conception*. Here is one of the earliest clear statements of the conception.

> This concept of set … according to which a set is something obtainable from the integers (or some other well-defined objects) by iterated application of the operation "set of," not something obtained by dividing the totality of all existing things into two categories, has never led to any antinomy whatsoever. (Gödel, 1964, pp. 474–75)

10.3 FROM TYPE THEORY TO THE ITERATIVE CONCEPTION

Let me digress briefly to describe a related insight of Gödel's, namely that the type-theoretic and set-theoretic schools have far more in common than initially meets the eye. Gödel goes so far as to claim that set theory is "nothing else but a natural generalization of the theory of types, or rather, it is what becomes

143

of the theory of types if certain superfluous restrictions are removed" (1933, pp. 45–46).[7]

One of the "superfluous restrictions" that Gödel has in mind concerns type theory's strict division into layers or types. As we have seen, type theory cannot allow an individual to be a member of a clan directly but only via some family (cf. §9.3). Set theory takes a far more relaxed approach by allowing an object to be a member of a set of any higher level, not just from the level immediately above. For example, the prohibition against an individual being a member of a clan is lifted.

Another difference between the two approaches concerns how the levels are represented. While type theory represents the levels by means of syntactic types, set theory adopts an untyped language and lets the levels be ontological in character. Thus, instead of using a typed variable, such as x^2, to represent a clan, set theory uses an untyped variable but says of its value that it was "formed" only at stage 2. Striking though this difference may seem, Gödel regards it too as superficial. Regardless of how the levels are represented, they are present on both approaches and play a crucial role in blocking the set-theoretic paradoxes. Consider Russell's paradox. Type theorists deny that the condition used to define the Russell set, namely non-self-membership, is so much as meaningful. Set theorists are more relaxed and allow every membership claim to be meaningful. As Gödel realized, any claim that the type theorists deem meaningless may instead be stipulated to be false. Thus, in set theory it is always false to say of a set of one level that it has elements from the same level or above. We enforce this idea by requiring that a set have members from lower levels only. This suffices to block Russell's paradox. The mentioned requirement means that *every* object satisfies the condition of non-self-membership. So there is no stage at which all the objects that satisfy this condition are formed and available to serve as members of a set. Hence, there is no Russell set. The other paradoxes are blocked in similar ways.

The final difference concerns the height of the two hierarchies. While type theory has syntactic levels indexed by all and only

[7] Linnebo and Rayo (2012) expands on the discussion in this section.

the natural numbers, set theory permits infinite ontological levels indexed by all the ordinal numbers. The process of set formation should be allowed to continue as far as possible.[8]

By holding that these three differences are superficial, Gödel in effect regards the Simple Theory of Types as a weak form of set theory, developed with a needlessly complicated syntax and subject to gratuitous restrictions, such as the ban on individuals being members of what Russell calls "clans."

10.4 STAGE THEORY

The first of the three questions prompted by the iterative conception asks which axioms of ZFC the conception motivates. The question is famously discussed in Boolos (1971). First, Boolos attempts to make the iterative conception formally precise by formulating an axiomatic theory aimed to describe how sets are formed in stages. Then, he proceeds to examine which axioms of ZFC this "stage theory" supports.

The stage theory is formulated in a language with separate systems of variables for stages and sets. There are three atomic predicates. In addition to the usual membership predicate '$\in$,' we write '$s < t$' for 'stage s is earlier than stage t' and 'Fxs' for 'the set x is formed at stage s.' The theory itself begins by describing the structure of the stages. First, the relation $<$ of 'earlier than' is a strict linear order.[9] Next, there is an initial stage, and immediately after any stage there is another. Finally, we lay down that there is a limit stage, that is, a stage other than the initial one that is not immediately after another stage.

The theory proceeds to describe how the sets are formed in stages. Every set is formed at some unique stage. Next, because the elements of a set have to be available before the set is formed, we lay down that each element of a set has to be formed at an

---

[8] Notice that this liberalization would not have been possible with the type theorists' strict layering, as there can be no level immediately below ω.

[9] That is, $<$ is irreflexive, transitive, and trichotomous (i.e., for any x and y, we have $x < y, x = y$, or $y < x$).

earlier stage than the set itself. Once the elements of a set are available, however, we immediately form the set. Thus, we lay down that every set is formed at the first stage which is after the stages at which its elements are formed. Next, we would like to say something about which objects are available at some stage. The answer is obvious. For any condition φ and stage s, the things that satisfy φ and are formed before s are available. Thus, we lay down that at any stage there is a set of all and only those things that satisfy φ and are formed before that stage. Finally, we would like to adopt a closure condition to the effect that there are no sets other than those formed in the aforementioned ways. We can do this by adopting induction axioms for the membership relation. That is, if some condition is satisfied by every non-set and by every set whose elements satisfy the condition, then every object whatsoever satisfies the condition.

We are now ready to inquire which axioms of ZFC are entailed by this explication of the iterative conception. It turns out that many are. The Powerset axiom provides a good illustration. Consider a set z, which must be formed at some stage s. Every element of z must therefore be formed before s. This also ensures that every element of every subset of z is formed before s. It follows that every subset of z is formed at or before s. But this means that all the subsets of z are available at s and that the desired powerset of z therefore is formed at the next stage. In fact, it turns out that all but three of the axioms of ZFC can be proved by this style of reasoning.

One of the exceptions is Extensionality. But Boolos regards this axiom as "quasi-analytic" and therefore unproblematic. He finds it more worrisome that neither Replacement nor Choice follows from the stage theory. Should we be worried? To establish that these axioms lack support from the iterative conception, it would first have to be argued that Boolos's stage theory captures the full content of the iterative conception. It is far from obvious that it does. Given the emphasis that the iterative conception places on the arbitrary nature of the set formation, it is not implausible to add a version of Choice to the stage theory. As Boolos shows, the ordinary axiom of Choice can then be derived. An analogous approach is possible with regard to Replacement, which can be derived if we make sufficiently strong assumptions about how many stages there are. However, this observation falls short of an unconditional answer to the question of whether Replacement follows from the iterative conception.

In my view, this shortcoming highlights an important weakness of Boolos's stage theory, namely that it fails to provide an independent handle on how many stages there are. The stages make up the spine along which the procedure of set formation unfolds. But on Boolos's account this spine is simply provided "from the outside." If we assume a stage-theoretic version of the axiom of Infinity, the corresponding set-theoretic axiom follows; and likewise for Replacement. This means that *neither* axiom receives any genuinely independent support from Boolos's stage theory.

Can we do better? Might the iterative conception shed light on the length of the spine? Recall the idea that the process of set formation should extend "as far as possible." Gödel had high hopes for a subtle explication of this idea in terms of the universe of sets being *indistinguishable* from its initial segments. According to this explication, any property that might distinguish the universe fails to do so as it is also had by one of its initial segment. This motivates a so-called *reflection principle*:[10]

$$\forall \vec{x} \, \exists \alpha \, \forall \vec{y} \in V_\alpha \left(\varphi(\vec{x}, \vec{y}) \leftrightarrow \varphi(\vec{x}, \vec{y})^{V_\alpha} \right)$$

A verbal gloss may be useful. The idea is that, for any condition φ utilizing parameters $\vec{x}$, there is an ordinal α such that the universe and the initial segment V_α look alike with respect to this condition. Thus, the condition fails to distinguish the universe from one of its initial segments. It is a pleasing fact that this reflection principle entails both the axioms of Infinity and Replacement. So if Gödel is right that the principle is part of the iterative conception, then this conception will motivate more axioms than Boolos thought, and in fact do so in a less circular way.

---

[10] As usual, $\vec{v}$ abbreviates a finite string of variables. And φ^{V_α} is the result of restricting all quantifiers in φ to V_α; e.g., '$\forall x$' is replaced by '$(\forall x \in V_\alpha)$.'

10.5 Understanding the Generative Vocabulary

Our second question is how to understand the temporal and constructive vocabulary that is used to describe the iterative conception. This question receives an influential discussion in Parsons (1977).

Interpreted literally, the usual explanation of the iterative conception suggests a view akin to that of constructive mathematics. Perhaps sets are "formed" through a synthetic operation of the mind that brings such objects into existence. This interpretation is deeply problematic, however. There are of course the general misgivings about mathematical constructivism discussed in §5.2. But the application of constructivist ideas to set theory is particularly problematic because of the daunting size of the process and the objects in question. As Parsons observes, ordinary time is simply not rich enough to provide a spine along which the process of constructing sets could take place. This process would require a "super-time," and we would need a "super-mind" to carry out the constructions in this time. These extreme idealizations would make the analogy with constructive mathematics far less apt.

It might be better simply to dismiss the constructive language of the iterative conception as mere rhetorical flourishes, which can be eliminated without any real loss of content. One proponent of this view is Boolos (1989), who suggests that the real content of the iterative conception is summed up in the observation that the cumulative hierarchy V can be retrieved as the union of the ranks V_α for each ordinal α. This observation is just a theorem of ZFC, which in turn is formulated in the austere language of ordinary set theory, without any of the problematic vocabulary. As we shall see shortly, however, it is doubtful that this minimalist version of the iterative conception can deliver all that the conception promises.

A third answer, due to Parsons, is to give the iterative conception a more ontological explication. According to him, "[w]hat we need to do is to replace the language of time and activity by the more bloodless language of potentiality and actuality" (1977, p. 293). Here is a useful summary of his proposal:

A multiplicity of objects that exist together *can* constitute a set, but it is not necessary that they *do*. ... However, the converse does hold and is expressed by the principle that the existence of a set implies that of all its elements. (pp. 293–94)

This requires some explanation. First, there is the idea that a set exists *potentially* relative to its elements. When the elements of some would-be set exist, we have all that is needed to define or specify the set in question: it is the set of precisely *these things*. Then, there is the related idea that the elements are ontologically prior to their set. *They* can exist although *it* does not—much like a floor of a building can exist without the higher floors that it supports. But a higher floor cannot exist without the lower floors that support it. Likewise, a set cannot exist without its elements, which are prior to it and on which the set is therefore ontologically dependent.

Clearly, this ontological explication of the iterative conception takes us deep into metaphysics. But this need not be problematic. The relevant metaphysical ideas have a fairly strong intuitive basis, and they give rise to a theory with substantial explanatory power—as we shall see in the remainder of this section and in the next, respectively. Let us begin with the idea of ontological dependence. Consider some things and one of these things. These things could not have existed without this one member. *They* are ontologically dependent on *it*. If the one member was destroyed, for example, then the remaining things would be distinct from the things with which we began; for some things cannot be identical with our initial things unless they have the very same members. Let us now return to the case of sets. Since a set is the result of applying the "set of" operation to some things, it is plausible to take the relation of ontological dependence to obtain also between the set and each of its elements.

Next, there is the idea of a set as merely potential relative to its members. This is implicit in Gödel's talk of a "set of" operation that is applicable to any "well-defined objects." *First* we pin down some "well-defined objects." *Then* we apply the "set of" operation to these objects. Applying this operation corresponds

149

to a permissible mathematical definition. For whenever we have pinned down some "well-defined objects," this suffices to define their set. This raises a concern, however. Why doesn't Gödel's "set of" operation lead to naive set theory, which is inconsistent? The concern is nicely articulated by Jonathan Lear:

> There are two beliefs associated with the iterative conception of set that are apparently mutually inconsistent: (i) Given any well-determined objects, they can be collected together into a set by an application of the *set of* operation. (ii) There is no set of all sets. (1977, p. 86)

The iterative conception is based on repeated application of the "set of" operation, which can be applied to any "well-determined objects." Simultaneously, the iterative conception purports to describe the cumulative hierarchy that results from this repeated application. So presumably there are some "well-determined objects" that make up the hierarchy. We should therefore be able to apply the "set of" operation to these objects as well. But doing so would yield a universal set, which is prohibited on the iterative conception.

10.6 ACTUALISM VERSUS POTENTIALISM

In order to respond to the threat of paradox, we need to distinguish between two different philosophical orientations toward set theory.

On a traditional platonist conception, mathematics is concerned with a fixed and determinate universe of abstract objects. Indeed, this is part of the analogy between mathematics and the empirical sciences on which platonism is based (cf. §1.4). Just as astronomy, say, is concerned with a fixed and determinate universe of stars, galaxies, and gas clouds, so mathematics is concerned with its own fixed and determinate universe of numbers, sets, and spaces. Other than lacking spatiotemporal location and being causally inefficacious, the objects of mathematics exist in the same way as those of astronomy and are all "available" to be talked about and quantified over in the same straightforward and unproblematic way as stars and galaxies. Let us call this orientation *actualism.*

Other approaches to mathematics reject the actualist conception of its subject matter. It is impossible to ascribe to mathematics a fixed and determinate universe of mathematical objects. For any such universe could be used to define an even larger such universe, which would be no less legitimate or mathematically interesting than the previous one. One such alternative approach derives from the ancient idea of potential infinity, according to which there are mathematical operations that can be applied indefinitely but whose applications can never be completed (cf. §§4.4 and 5.4). Zermelo famously defended an analogous view of set theory. Although set theory abounds with actual infinities, a form of *incompletability* still remains. For any domain of sets—or indeed model of the axioms of set theory—can be extended to an even richer such domain or model:

> What appears as an 'ultrafinite non- or super-set' in one model is, in the succeeding model, a perfectly good, valid set with both a cardinal number and an ordinal type, and is itself a foundation stone for the construction of a new domain. (Zermelo, 1930, p. 1233)

That is, the hierarchy of sets is "open-ended" and incapable of being completed. Similar ideas are found in the ontological explication of the iterative conception, which we discussed in the previous section. Let us call this orientation *potentialism.*

Given the potentialists' acceptance of actually infinite sets, one may wonder whether the two orientations differ in substance, not just in their choice of imagery and heuristics. In fact, an interesting difference emerges when we return to the threat to the iterative conception articulated by Lear and broached in the previous section. It turns out that potentialists have a response to this threat that is not available to actualists. Recall that Gödel says the "set of" operation can be applied to any "well-defined objects"; other expositors require that the objects be "available" or "given." What do these requirements amount to, however? Potentialists can offer an appealing explication. To be "available" or "given" is to *possibly co-exist* and thus to be capable of

151

completion.[11] As Zermelo observed, the cumulative hierarchy of sets is not "available" or completable in this way. This assuages Lear's concern. Since the cumulative hierarchy lacks the kind of well-definedness that is required for the "set of" operation to be applicable, the push toward a universal set is blocked and the iterative conception is coherent after all.

Things look very different from an actualist point of view. It is part and parcel of this view that *all* sets are "available" or "exist together." This deprives actualists of the proposed response to Lear's concern. In fact, since actualists accept pluralities—or "finished" multiplicities—that don't form sets, they are vulnerable to the charge that they have arbitrarily truncated the cumulative hierarchy at some point where it might have been continued to even higher reaches.

Our discussion has made heavy use of modal talk. So it is reassuring that all of our distinctions and arguments can be made explicit using the resources of modal logic. At the heart of this modal explication is a principle concerning the existence of sets.

> (6) Necessarily, given any things, it is possible that these things form a set.

We may think of this as a potentialist version of the naive principle of set comprehension (cf. §4.2). The paradoxes show that the actualist version of the principle is unacceptable: it is disastrous to assume that any things *do* form a set. But the potentialist version states only that any things *may* form a set. This principle is consistent, intuitive, and theoretically useful.

We also need a principle of extensionality that specifies how sets are individuated. Here is a plausible option:

---

[11] See Parsons (1977) and Linnebo (2010). In fact, this answer harks back to Cantor's famous distinction between "consistent" and "inconsistent" multiplicities. Since only the former "can be thought of as *finished*," only they give rise to sets.

(7) Necessarily, if x is the set of some things uu and y is the set of some things vv, then $x = y$ iff $uu = vv$.

There are other natural ways as well to strengthen the potentialist set theory. It is, for example, natural to require that a set have the same elements at every possible world at which it exists; that the stages of the set formation be well-founded; and that the subsets of a set be formed no later than the set itself. Properly explicated, these additions turn out to justify Zermelo set theory minus the axiom of Infinity, much like Boolos's stage theory justifies a similar amount of set theory. Finally, it is natural to add a reflection principle in order to capture the idea that the process of set formation is continued as far as possible. Doing so allows us to justify the axioms of Infinity and Replacement as well, and thus all of ZF.[12]

Summing up, the modal explication of the iterative conception provides an interesting alternative to Boolos's stage theory. We have discussed three attractive features of this alternative. It puts the potentialist conception of the hierarchy to explanatory use. It makes available a promising response to Lear's concern. And it shows that naive set theory contains valuable ideas—provided that these are developed in a potentialist rather than actualist setting.

SELECTED FURTHER READING

Boolos (1971) develops the stage-theoretic exposition of the iterative conception. Paseau (2007) discusses which ZFC axioms are justified by this exposition. Parsons (1977) is another classic analysis of the iterative conception, while Potter (2004) is a useful and quite accessible book-length discussion. An earlier, but still very valuable, discussion of the ideas and extrapolations associated with the iterative conception is Bernays (1935).

---

[12] See Linnebo (2013b) for details.

Structuralism

11.1 MATHEMATICS AS CONCERNED WITH ABSTRACT STRUCTURE

Structuralism is a philosophical view that emphasizes mathematics' concern with abstract structures, as opposed to particular systems of objects and relations that realize these structures. Consider three children linearly ordered by age and three rocks linearly ordered by mass. These two systems of objects and relations realize the same abstract structure, namely that of three objects in a linear order. All that matters for mathematical purposes, according to structuralism, is the abstract structure of some system of objects and relations, not the particular natures of these objects and relations.

This strong emphasis on abstract structure is not uncontroversial. According to the iterative conception of sets, for example, set theory has as its subject matter the cumulative hierarchy of sets. This is a *particular* system of objects, ordered by the *particular* relation of membership. While it is uncontroversial that children, rocks, age, and mass are of no particular concern to mathematics, it is less clear what to say about sets and the membership relation as understood on the iterative conception. We shall examine whether structuralism is merely a methodological supplement to set theory or whether there is a genuine tension between the two.

A methodological form of structuralism emerged in nineteenth-century mathematics (cf. §3.4). Prior to that, mathematical notions and theories tended to have very particular interpretations; for example, geometry was about physical space, arithmetic about counting, and set theory about collecting objects. In the course of the nineteenth century, such particular interpretations became less important. Geometry provides a striking example. Following the discovery of non-Euclidean geometry, the link between geometry and physical space was abandoned in favor of Hilbert's more abstract approach, which regards geometry as the study of any system of objects structured in some appropriate and loosely "spacelike" manner. An analogous development took place in algebra, where theories of algebraic structures such as groups, rings, and fields were formulated with the explicit aim of *not* having a particular interpretation. The aim was instead to characterize some important classes of structures that have multiple realizations throughout mathematics and perhaps also in the physical world. As a final example, consider the "arithmetization" of analysis, culminating in Dedekind's entirely abstract characterization of the structure of the real number line, which was shown to have multiple realizations.[1]

The philosophical form of structuralism goes beyond the methodological one by holding that mathematics is the study of abstract structures—or *patterns*, as they are often called. The view is nicely encapsulated in the claim that "[m]athematics is simply the catalogue of all possible patterns" (Barrow, 2010).[2] Philosophical structuralism gains some of its credibility from the structuralist methodology that has come to dominate in mathematics. It also promises progress on various philosophical questions. A nice example is the question of why abstract mathematics is applicable to the physical world. If mathematics is simply the study of "all possible patterns," as Barrow contends, "it is inevitable that the world is described by mathematics": for whatever abstract patterns that the world instantiates (or approximates) belong to mathematics, thus understood.

Other philosophical questions require structuralists to clarify their talk about abstract structures. Recall that two systems are said to have (or instantiate, or realize) the same abstract

---

[1] For the cognoscenti: The reals are characterized as a complete ordered field, and two famous realizations are provided by Dedekind cuts and equivalence classes of Cauchy sequences.

[2] Although John Barrow is a mathematical physicist, this is a philosophical claim about the nature of mathematics.

structure just in case they are isomorphic (cf. §3.4). We left open whether this talk of abstract structures should be understood as ontologically committing or merely as an innocent manner of speaking. This choice has resulted in a great schism within structuralism. *Eliminative structuralists* admit that such talk is heuristically useful but insist that it be understood so as to avoid genuine commitment to abstract structures or patterns. *Noneliminative structuralists* disagree and find it appropriate to undertake such commitments. As we shall see, both movements trace their roots back to Richard Dedekind (1831–1916), who can thus be regarded as the father of mathematical structuralism.

11.2 ELIMINATIVE STRUCTURALISM

How can talk about abstract structures be understood as non-committal? Reflection on the ancient philosophical debate about the ontological status of universals suggests an answer. We often speak as if universals exist, for example, when we say that two objects instantiate the same color. But opponents of universals understand this as just stating that the objects are chromatically equivalent. Eliminative structuralists adopt an analogous strategy. When we informally say that some systems instantiate the same abstract structure, this should be understood as just stating that the systems are isomorphic.[3]

So far, so good. But how can this be turned into a structuralist account of mathematics? Consider the case of arithmetic. On a traditional platonistic interpretation, arithmetic is about particular objects, each with a particular nature. For example, 1 is the number of any collection with a unique member, and 2 is the number of any pair-collection. These objects stand in various

---

[3] Harking back to the ancient debate about the ontological status of universals, eliminative structuralism is also known as *in re* structuralism, and the noneliminative, as *ante rem*; see Shapiro (1997). Following Parsons (1990), I prefer the labels "noneliminative" and "eliminative," which carry fewer connotations from the ancient debate.

arithmetical relations. It turns out to be possible to describe all these relations in terms of just the successor function s, which maps each number n to its successor $n + 1$. Let $\mathbb{N}(x)$ mean that x is a natural number. We now formulate a version, PA^2, of second-order Dedekind-Peano arithmetic.[4]

(1) $\mathbb{N}(0)$

(2) $s(x) \neq 0$

(3) $s(x) = s(y) \to x = y$

(4) $\forall X(X(0) \wedge \forall y(X(y) \to X(s(y)))) \to \forall y(\mathbb{N}(y) \to X(y)))$

That is, 0 is a natural number; 0 is not the successor of anything; the successor function s is one-to-one; and finally, any collection X that contains 0 and is closed under s contains all the natural numbers. This last axiom, known as the *induction axiom*, ensures that $\mathbb{N}$ is *the smallest* collection that contains 0 and is closed under s. Since the axiom quantifies over collections, the axiomatization counts as second-order.

(Let me digress briefly to explain how *first-order Dedekind-Peano arithmetic (PA)* differs from PA^2. Since PA uses only first-order logic, the induction axiom is replaced by an *induction scheme*, every instance of which is an axiom:

$$\varphi(0) \wedge \forall y(\varphi(y) \to \varphi(s(y))) \to \forall y(\mathbb{N}(y) \to \varphi(y)))$$

Next, the language of PA adds two-place function symbols + and · for addition and multiplication, respectively, as well as axioms providing the usual recursive characterization of these two functions:

$$x + 0 = x \qquad\qquad x \cdot 0 = x$$

$$x + s(y) = s(x + y) \qquad x \cdot s(y) = x \cdot y + x$$

This concludes the digression.)

---

[4] This version differs from that of §2.6 by invoking a successor *function* rather than a *relation* of immediate predecession. The difference is mathematically unimportant.

Dedekind showed how to develop a more abstract approach to arithmetic. We begin by characterizing the relevant abstract structure. Consider a collection of objects X with a designated member a and a function f. We would like to state that this system has the abstract structure described by the theory PA^2. This is easily achieved by means of the conjunction of the axioms—but with X, a, and f replacing $\mathbb{N}$, 0, and s, respectively. Let $PA^2[X, a, f]$ abbreviate this conjunction. We now say that a system $\langle X, a, f \rangle$ is *simply infinite* just in case $PA^2[X, a, f]$; in other words, just in case PA^2 holds of the system where X, a, and f play the roles of $\mathbb{N}$, 0, and s, respectively.

Equipped with this definition, Dedekind (1888) proved his famous *categoricity theorem*, which states that any two simply infinite systems are isomorphic.[5] This important result means that the definition of a simply infinite system succeeds in uniquely characterizing the intended abstract structure, namely that of the "true" natural number system $\langle \mathbb{N}, 0, s \rangle$. More generally, a theory is said to be *categorical* when any two systems satisfying its axioms are isomorphic. Another famous example of a categorical theory is that of a complete ordered field, which uniquely characterizes the structure of the real number line. We shall continue to focus on the case of arithmetic.

Simply infinite systems can obviously differ vastly in their internal constitution. The system can be based on a sequence of stars, spacetime points, sets, or whatever. Since any two simply infinite systems are isomorphic, however, they share the same *structural properties*. The objects are linearly ordered, for example, and this order has a unique initial object but no terminal object. To make this precise, consider any sentence φ of the language of PA^2. Let $\varphi[X, a, f]$ be the result of substituting X, a, and f for $\mathbb{N}$, 0, and s, respectively; intuitively, this new sentence

---

[5] The proof rests on a simple idea. Let $\langle X, a, f \rangle$ and $\langle Y, b, g \rangle$ be simply infinite systems. We define an isomorphism $\varphi : X \to Y$ by mapping one initial object, a, to the other, b, and then stepwise extending φ by letting $\varphi(f(x))$ be $g(\varphi(x))$ for any x in X. We use the induction axiom (iii) to show that φ is defined on all of X and is an isomorphism.

says of the system $\langle X, a, f \rangle$ what φ says of the "true" natural number system $\langle \mathbb{N}, 0, s \rangle$. We can now prove the following.[6]

Assume $\langle X, a, f \rangle$ and $\langle Y, b, g \rangle$ are simply infinite systems. Then $\varphi[X, a, f]$ iff $\varphi[Y, b, g]$.

This result opens an exciting possibility. So long as we are only considering structural properties, in the above sense, any simply infinite system is as good as any other: they all yield the same answers. There is thus no need to consult the platonist's preferred system of "true" natural numbers. This system is exposed as an idle wheel. A fully structuralist alternative is possible.

To understand the alternative better, consider $1 + 1 = 2$, which can be understood as concerned, not with *particular* objects, but with a *general* fact about simply infinite systems, namely that objects occupying certain positions bear a certain relation to one another. As far as arithmetic is concerned, the nature of the objects occupying the positions is irrelevant; what matters is only the abstract structure of the system to which the objects belong. More generally, a sentence φ of the language of PA^2 can be analyzed as:

$$(11.1) \qquad \forall X \forall a \forall f \left(\text{PA}^2[X, a, f] \to \varphi[X, a, f] \right)$$

This analysis works because φ is true of *one* simply infinite system—say the platonist's "true" natural numbers—just in case it is true of *all* such systems.

There is one catch, however. The proposed analysis assumes that there is *at least one* simply infinite system; otherwise, all instances of (11.1) would be vacuously true, with the disastrous result that both φ and $\neg \varphi$ are translated as truths. This is a version of what we called the problem of model existence (cf. §3.5).

Are we entitled to assume there are simply infinite systems? One option is to appeal to set theory for realizations of this and other abstract structures. The resulting view is known as *set-theoretic structuralism*. The set-theoretic axioms are regarded

---

[6] The proof goes by induction on the syntactic complexity of φ.

as foundational (as making assertions), and all other axioms, as structural (as parts of definitions; cf. §3.5). This is a powerful view, as evidenced by the fact that this is now fairly standard mathematical methodology. The view does, however, rely heavily on set theory, which requires an independent, nonstructuralist account, perhaps of the sort discussed in the previous chapter.

Can the reliance on set theory be avoided? The question is particularly important to aspiring nominalists, who need to solve the model existence problem without appealing to abstract objects. A nominalist will begin by reminding us that the models in question need not be set-theoretic but can be characterized using the resources of second-order logic (cf. §3.4). Even so, however, it is not clear that the desired models can be found. As Hilbert observed, we have no guarantee that there are infinitely many concrete objects (cf. §4.3).

A promising option is known as *modal structuralism*.[7] Consider the case of arithmetic. The idea is that it suffices that there *could be* a simply infinite system. Even if Hilbert is right about the actual universe, this weaker modal assumption seems plausible. There could, for example, be a simply infinite system of stars. Modal structuralism next shows that this weaker assumption suffices for an account of arithmetic. The key is to translate an arithmetical sentence φ as *the necessitation* of (11.1), thus ensuring that the translation talks not just about actual but also about possible simply infinite systems.[8] From a technical point of view at least, the approach works. This suggests that it might be possible to eliminate abstract mathematical objects— beginning with the natural numbers—in favor of modal claims about possibility and necessity.

Would this elimination of abstract objects in favor of modal claims constitute progress? Are the modal claims less philosophically problematic than the corresponding ontological claims? A proper discussion of these controversial questions would take us far beyond the philosophy of mathematics. What *is* clear is that modal structuralism is a philosophically interesting competitor

---

[7] See Hellman (1989).

[8] The necessitation of ψ is the claim that ψ is necessary.

to platonism. In particular, the view suggests that mathematical objectivity need not be underpinned by the existence of mathematical objects. On the modal structuralist account, each arithmetical statement has an objective, mind-independent truth-value. The view is thus a version of *truth-value realism*. But this realism is achieved without any commitment to mathematical objects. Modal structuralism thus avoids *object realism*. This combination of truth-value realism with object anti-realism is made possible by adopting a nonclassical semantics for the language of arithmetic, based on the necessitation of the analysis in (11.1).

11.3 NONELIMINATIVE STRUCTURALISM

Contemporary mathematics often professes to adhere to set-theoretic structuralism. But in practice, many mathematicians continue to talk about abstract structures as if these were objects, such as "the real numbers" or "the cyclic group of order 4," as well as about individual mathematical objects, such as the number 0 and the identity element of the mentioned group. Noneliminative structuralists believe this kind of talk should be taken seriously.[9] There really are abstract structures, and ordinary mathematical objects are merely positions in these structures. Noneliminative structuralists typically add that abstract structures exist in their own right, not in virtue of their particular instantiations. This would mean that there is no problem of model existence. For example, the abstract structure described by arithmetic exists in its own right, independently of any prior realization by some particular simply infinite system.

What exactly are the abstract structures postulated by noneliminative structuralists, and how should their objects or positions be understood? It is particularly important to understand how this form of structuralism differs from traditional platonism. The usual answer is two-pronged. One distinguishing

[9] See, e.g., Resnik (1981), Parsons (1990), and Shapiro (1997).

feature of mathematical objects is said to be their "incompleteness" with respect to their properties. Consider physical objects, which have intrinsic properties, such as mass, shape, and chemical composition; many of them have an internal composition; and they stand in (fairly) determinate relations of identity and distinctness. Noneliminative structuralists take mathematical objects to be very different. Michael Resnik provides a widely endorsed statement of the view:

> In mathematics, I claim, we do not have objects with an 'internal' composition arranged in structures, we have only structures. The objects of mathematics [...] are structureless points or positions in structures. As positions in structures, they have no identity or features outside a structure. (1981, p. 530)

According to this *incompleteness claim*, as I shall call it, mathematical objects differ from physical objects by having no internal nature or composition, and no nonstructural properties. In particular, cross-structural identity statements involving mathematical objects have no objective truth-values. There is, for example, no fact of the matter about whether the natural number zero is identical with the empty set.

The incompleteness claim has its roots in Dedekind's work. Consider the simply infinite system comprising the finite von Neumann ordinals: $\varnothing, \{\varnothing\}, \{\varnothing, \{\varnothing\}\}, \ldots$. These ordinals can obviously *play the role of* the natural numbers and are indeed often made to do so. Yet it seems perverse to *identify* the natural numbers with the ordinals and thus conclude, for example, that 0 is an element of 1! Numbers stand in arithmetical relations, not in set-theoretic ones. Moved by such considerations, Dedekind wishes to purge mathematical object of all such "foreign properties" (1963, p. 10). So he proposes that the natural numbers be obtained by erasing "the special character" of the elements of some given simply infinite system, retaining only their purely structural relations to one another:

> If in the consideration of a simply infinite system X set in order by a transformation f we entirely neglect the special character of the elements; simply retaining their distinguishability and taking into

account only the relations to one another in which they are placed by the order-setting transformation f, then are these elements called natural numbers. (1888, §73)

These ideas conflict with a traditional platonistic outlook, as expressed, for instance, when Russell writes of the natural numbers that "[i]f they are to be anything at all, they must be intrinsically something; they must differ from other entities as points from instants, or colours from sounds" (1903, p. 249).

The incompleteness claim is deeply problematic, however. Natural numbers appear to have a wide variety of nonstructural properties, such as being abstract, being a natural number, being the number of planets, and being Dedekind's favorite number. Moreover, many cross-category identities appear to be false, not indeterminate. Since 0 is a natural number but $\varnothing$ is not, it is immediate from Leibniz's law that they are distinct. Indeed, if mathematical objects are just positions in patterns, it must presumably be a property of each such object to be a position in the particular pattern to which it belongs.

The most one can hope for is that some qualified version of the incompleteness claim might be acceptable. Although mathematical objects do have nonstructural properties, perhaps all of their properties *from some important class* are purely structural. Provided that the counterexamples just discussed lie outside of the mentioned class, this qualified claim might still salvage some structuralist insight. For this response to be acceptable, however, the structuralists need to circumscribe the mentioned class of properties—and to do so in a way that makes the resulting view both interesting and true. This is not an easy task.[10]

The second distinguishing feature of mathematical objects, according to noneliminative structuralists, concerns ontological dependence (cf. §10.5). Traditional platonism compares mathematical objects with physical objects. Both are said to exist independently of cognizers and typically also of one another. By contrast, noneliminative structuralists maintain that mathe-

---

[10] See Shapiro (2006), but also Linnebo (2008), which expands on the present section.

matical objects from the same structure have a special kind of dependency on each other and on their structure:

> The number 2 is no more and no less than the second position in the natural number structure; and 6 is the sixth position. Neither of them has any independence from the structure in which they are positions, and as positions in this structure, neither number is independent of the other. (Shapiro, 2000, p. 258)

According to this *dependence claim*, as I shall call it, the "essence" of a natural number lies in its relations to other natural numbers. As a mere position in this structure, it would not exist without the structure; nor could the structure exist without its other positions. This is an attractive explication of the non-eliminative structuralist view that mathematical objects are mere positions in patterns.

Is the dependence claim true, however? My own view is that the claim has many true instances, but there are also counterexamples. On the iterative conception of sets, for example, each set depends on its elements but not on every other set; in particular, not on sets formed at the same or later stages (cf. §10.5).

11.4 ABSTRACT STRUCTURES BY ABSTRACTION?

One way to make sense of abstract structures is by understanding how they are related to the particular systems that realize them. And it is natural to think that each abstract structure is related to the particular systems that realize it by some form of abstraction. So let us take a closer look at how abstraction might help our present investigation.

As we have seen, Dedekind emphasizes that the elements of each particular system must be purged of all their "foreign properties." We must, however, retain all of their structural relations to one another, thus ensuring that the abstract structure that results is isomorphic to the system with which we began. As Frege joked in a related context, we need a lye that is just strong enough to wash away all of the unwanted properties, while preserving all of the desired ones.[11] What might this metaphysical lye be? Dedekind at times answered that our minds "create" new objects that have been appropriately purified. Frege rightly protested that this answer is unhelpful. Whatever positions in abstract structures are, they are not in any obvious way created by us. Things that we create, such as buildings and contracts, did not exist prior to their creation, in sharp contrast to mathematical structures, which are naturally taken to exist in an atemporal way (cf. §1.3).[12]

Might Fregean abstraction be of any help? Some reason for optimism is provided by Shapiro's characterization of a structure as "the abstract form of a system." He also claims that an abstract structure stands to the systems that realize it the way that a type stands to its tokens.[13] This suggests an abstraction principle for structures: two systems instantiate the same abstract structure just in case they are isomorphic. However, the abstract structures are supposed to contain abstract positions, just as a system contains objects. Can the abstract positions too be obtained by Fregean abstraction? Consider objects a and a' from systems S and S', respectively. A natural suggestion is that the objects occupy the same abstract position just in case there is an isomorphism of S and S' that maps a to a'. This suggestion works well in many cases; for example, for any simply infinite system, we obtain pure positions that are themselves ordered as a simply infinite system: the 0 position, the 1 position, etc. But unfortunately, the suggestion has unacceptable consequences for

---

[11] Compare Frege (1894, p. 84).

[12] Dedekind's claims about our creation of mathematical objects might perhaps be interpreted as merely a clumsy way of saying that we create new languages for talking about mathematical objects. Consider the eliminative set-theoretic structuralism made available by Dedekind's categoricity theorem. As we have seen, this approach allows arithmetical truths to be stated in the ordinary arithmetical language—only that each such sentence is now interpreted as a generalization over simply infinite systems. *As far as this language is concerned*, it is indeed as if "purified" numbers have been created with no properties other than the structural ones expressible in the language of arithmetic.

[13] See Shapiro (1997, p. 74 and p. 84, respectively), as well as Linnebo and Pettigrew (2014) for discussion.

other structures. Consider, for example, the abstract structure of the "dumbbell graph":

$$\bullet \longleftrightarrow \bullet$$

Next, consider a particular system that realizes this abstract structure, say the one based on Romeo, Juliet, and their mutual love. Clearly, swapping Romeo and Juliet yields an isomorphic system. So the proposed individuation of abstract positions implies that Romeo and Juliet occupy one and the same abstract position! This, in turn, implies that the structure contains only one position, while in fact it contains two.

An alternative strategy is to understand an abstract structure as a *structured universal*, which is instantiated by all and only the members of some family of pairwise isomorphic systems. The abstract dumbbell graph, for example, might be the structured universal instantiated by the system consisting of Romeo, Juliet, and their mutual love, as well as by any isomorphic system. Thus, this structured universal has two argument places for objects and one for dyadic relations. This strategy requires that the argument places of a structured universal play two separate roles. First, since a structured universal can be instantiated by particular systems, its argument places must function as offices that can be occupied by appropriate entities. Additionally, the structured universal is supposed to be isomorphic to any system that realizes it. This requires that its argument places can also be construed as entities in their own right.[14] However, it is not clear how argument places can play both of these roles. Consider an argument place that is open to relations. This argument place must itself be a relation and thus capable of being predicated of other entities; otherwise the abstract structure could not be isomorphic to its realizing systems. This is a tall order. We understand what it is for an argument place to be *occupied* by some entity. What it would be for an argument place to be

---

[14] This dual role of the argument places corresponds to Shapiro's distinction between places as offices and places as objects. See Shapiro (1997, chap. 3). Consider the U.S. vice presidency. This office has had many occupants. But it can also figure as an object in its own right, for example, when we say that the vice president is the president of the Senate.

predicated of something is far less clear. An argument place is not the sort of thing we ordinarily regard as capable of being predicated of anything.

11.5 CATEGORY THEORY AND STRUCTURALISM

Category theory is an abstract algebraic tool developed around the middle of the previous century. One of its main purposes is to characterize mathematical structures and constructions only up to isomorphism. Although no proper explanation of category theory is possible here,[15] briefly comments on its significance for mathematical structuralism are appropriate. The basic idea is simple enough. Structuralism is the belief that all that matters in mathematics is preserved under isomorphism. And category theory characterizes things only up to isomorphism. So the two seem to be a perfect match.

To elaborate, let us consider an example of the kind of characterization that category theory makes possible. In orthodox set theory, the Cartesian product of two sets A and B is defined as the set of all and only those ordered pairs whose first member belongs to A and whose second member belongs to B—where ordered pairs in turn are a certain sort of set. Other mathematical structures have Cartesian products as well, for instance, differentiable manifolds and topological groups; their definitions require even more excruciating set-theoretic specificities. What is common to Cartesian products across all these structures? Category theory provides a beautiful and completely general answer in terms of what such products allow us to do, rather than in terms of the messy details of their inner make-up, which varies from case to case. A *Cartesian product* of A and B is, first, a set $A \times B$, equipped with projection mappings π_A and π_B from $A \times B$ to A and B, respectively. Second, the product has the following minimality property. For any pair of mappings f and g from C to A and B, respectively, there is a single map $f \times g$: $C \to A \times B$ such that $f = \pi_A \circ (f \times g)$ and $g = \pi_B \circ (f \times g)$;

and $f \times g$ is the *only* mapping that has this property.[16] We illustrate the situation using a commutative diagram:[17]

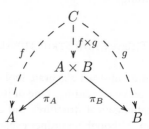

This characterization of Cartesian products is entirely in terms of their "functional role," not in terms of the messy details involved in the realization of this functional role. A Cartesian product of A and B is any structure with projections onto A and B, and which is minimal in the sense that any other structure with such projections has a unique map into the Cartesian product. This functional role characterization provides all we need to know about such products. For example, it implies that any two constructions that satisfy this role are isomorphic.

Many other mathematical constructions too can be characterized in terms of their functional role. This has been exploited to develop category-theoretic alternatives to standard set theory. In these alternatives, all constructions are characterized solely in terms of their functional roles, not in terms of their inner make-up. Its proponents claim this provides a better, and more structuralist, foundation for mathematics.

There can be no doubt that category theory is a powerful organizational tool, which enables us to do mathematics in a structuralist fashion in cases where we care about our constructions only "up to isomorphism." But what exactly are the items that are studied in this fashion? A conservative answer is that they are sets in the familiar old sense. If this is right, then the value

---

[16] $f \circ g$ is the composite function defined by $f \circ g(x) = f(g(x))$.

[17] A diagram is said to be *commutative* just in case any two journeys from one point to another give rise to identical functions.

of category theory is entirely of methodological nature. A competing, radical answer is that category theory studies its own *sui generis* objects with no more of a nature than what is expressed in their highly abstract category-theoretic characterizations. This would be a form of noneliminative structuralism.[18]

SELECTED FURTHER READING

Benacerraf (1965) is the classic defense of eliminative structuralism. Noneliminative structuralism is defended in articles by Resnik (1981) and Parsons (1990) and a monograph by Shapiro (1997). Hellman (2001) and Linnebo (2008) are critical discussions. Reck (2003) gives a useful exposition and analysis of structuralist ideas in Dedekind.

---

[18] See McLarty (2004).

CHAPTER TWELVE

The Quest for New Axioms

12.1 THE INCOMPLETENESS PHENOMENON

Does every meaningful mathematical question have an answer? Hilbert famously thought so, insisting that in mathematics there is no *ignorabimus*.[1] A good test case is *Cantor's continuum hypothesis* (henceforth, CH), which asserts that the cardinal number of the reals is the least cardinal number that is larger than that of the naturals. In other words, there is no cardinal number strictly between that of the naturals and that of the reals. Cantor tried hard to prove his hypothesis but failed. We now know why: the problem is independent of our standard set theory, ZFC. Assuming this theory is consistent, it *provably* does not settle CH one way or the other! The first half of this important meta-mathematical result was established in 1939, when Gödel proved that CH is consistent with ZFC. So ZFC doesn't prove the negation of CH (provided ZFC is consistent). Twenty-four years later, Paul Cohen established the second half by proving that the negation of CH is also consistent with ZFC. So ZFC doesn't prove CH either (provided ZFC is consistent). Many other independence results are known by now. But we shall use CH as our main example.

How should we respond to the independence of CH? One option is to throw up our hands and deny that the problem has an answer. But that would be rash. Compare Galois's famous result that we cannot trisect a given angle by means of ruler and compass constructions alone. This too is a mathematical result that places limits on our ability to solve certain

---

[1] The term is Latin for "we shall not know" and was a slogan of nineteenth-century thinkers who postulated strict limits to scientific knowledge.

mathematical problems. But it would be crazy to deny that angles have trisections just because these cannot be constructed by the mentioned means alone. Likewise, why shouldn't CH have an answer, even though ZFC alone cannot deliver it? This line of thought receives further support when we reflect on what CH says. Here is a particularly elementary formulation. *Every subset of the real numbers can be one-to-one correlated either with some natural numbers or with the real numbers themselves.* This seems a perfectly meaningful statement about mathematical structures that we know and accept, namely the naturals and the reals. So it seems that the statement must be either true or false— whether or not we are clever enough to determine which. These considerations encourage a bolder response to the independence result. Since the question has an answer, we must search for new axioms to supplement those of ZFC such that an answer can be proved.

These two responses to the independence of CH flow from two opposing philosophical views of set theory. Theorists of an antirealist bent believe that ZFC more or less exhausts our conception of sets and what there is to know in set theory. Where this authority is silent, there is simply no answer to be had. For example, Feferman (2014) contends that "CH in its ordinary reading is essentially indefinite (or 'inherently vague') because the concepts of arbitrary set and function needed for its formulation can't be sharpened without violating what those concepts are supposed to be about." Theorists of a stronger realist persuasion claim that the question of CH is posed in a meaningful language all of whose sentences have objective truth-values. So the question must have an answer, and the challenge is to find new axioms that will reveal it. In short, the different philosophical views give rise to different recommendations concerning mathematical practice. This is exciting. Our task of determining the limits of objectivity in mathematics bears on how mathematics should be practiced.

The greatest defender of the realist view is Gödel. So we shall investigate his influential, though controversial, view of mathematical evidence and of how new axioms might be defended.

12.2 Intuition and Intrinsic Evidence for Axioms

Gödel had great faith in our mathematical intuition. As we have seen, he goes so far as to claim that we "have something like a perception also of the objects of set theory, as is seen from the fact that the axioms force themselves on us as being true" (1964, pp. 483–84). If the axioms of ZFC have this ability to "force themselves on us," might not further axioms do so as well?

The answer will obviously depend on how mathematical intuition is understood. One option is to understand such intuition in a broadly perceptual manner. But then its reach will be far more limited than Gödel seems to have thought (cf. chap. 8). Thus understood, intuition can provide information about types that are either instantiated or can be imagined in a clear and distinct way, but not much more. An alternative is to understand mathematical intuition as a form of apprehension of conceptual truths. This is what Gödel has in mind in some of his writings. For example, he claims we need "a more profound analysis (than mathematics is accustomed to giving) of the meanings of the terms occurring in [CH]" (1964, p. 473). Evidence of this form is known as *intrinsic*, as it flows from our set-theoretic concepts alone. Gödel is particularly hopeful that a deeper analysis of the iterative conception of sets might yield evidence for new axioms. It is not hard to see why. The conception motivates many of the axioms of ZFC (cf. chap. 10). So it might well be capable of motivating further axioms as well.

To explore the prospects for intrinsic evidence, let us first recall another form of incompleteness, namely that of the Gödel sentence of one of our foundational theories (cf. §4.6). This sentence too is left undecided by the theory in question. As Gödel was well aware, however, the Gödel sentence can be decided in a perfectly natural way by adopting a stronger theory, which accepts arbitrary sets of the objects with which this original theory is concerned. This is an entirely natural extension: we just consider one layer of sets on top of the objects described by the original theory. In this extension, the Gödel sentence of the original theory can be *proved*. A question that was left open by the original theory can thus be answered in an entirely natural and nonarbitrary manner. The answer we get is unsurprising. What the Gödel sentence "says" is true: it cannot be proved in the original theory. What *is* surprising is that this answer can be proved simply by considering one layer of sets on top of the objects with which we began.

The independence of CH is far more stubborn than that of a Gödel sentence. It is not sufficient just to "add" one layer of sets. Gödel is nevertheless hopeful that we might make progress by adding *a vast number* of such layers. We should apply the operation "set of," which takes us from one level of the cumulative hierarchy to the next, many more times than what is required by ZFC alone. Since Gödel takes it to be part of the iterative conception that this operation should be applied *as many times as possible*, he believes that such extensions are motivated by the iterative conception.

> [T]he very concept of set on which [the current axioms] are based suggests their extension by new axioms which assert the existence of still further iterations of the operation of "set of." These axioms can be formulated also as propositions asserting the existence of very great cardinal numbers. (1964, p. 476)

In this way, Gödel thinks, *large cardinal axioms* can be seen as "only unfold[ing] the concept of set" (ibid., p. 477).

Gödel's large cardinal program has borne many fruits. Unfortunately, it has not resolved CH, which appears to be independent of every extension of ZFC by large cardinal axioms (provided the extension is consistent).[2] To settle CH, we appear to need some different form of axioms.

12.3 EXTRINSIC EVIDENCE FOR AXIOMS

While disappointing, this later development would not have stopped Gödel. For he postulates another form of evidence that is more indirect, but also more far reaching, than the intrinsic evidence discussed so far. To explain this *extrinsic* form of

---

[2] See Koellner (2013).

evidence, it is useful first to recall the traditional conception of mathematical justification, especially in the Euclidean and rationalist tradition. Here mathematical axioms are regarded as self-evident and epistemologically fundamental. Gödel's notion of extrinsic evidence completely abandons these requirements.

Gödel was not the first to move in this direction. The traditional conception of axioms came under increasing pressure throughout the nineteenth century, culminating in two influential (and nearly contemporary) attacks. As we have seen, Zermelo (1908) supplements traditional appeals to self-evidence with considerations about the needs of well-established branches of mathematics. When an axiom is indispensable to some established branch of mathematics, this is taken to support the axiom. The other attack is initiated by Russell (1907), who defends a "regressive method" in mathematics. He abandons the requirement that axioms be self-evident and emphasizes instead how axioms may be less obvious than the proposition they entail. In what sense, then, are the non-self-evident propositions still axioms? Russell urges us to look to the empirical sciences. The fundamental principles of these sciences are far less obvious than their consequences concerning the observable, but they are nevertheless justified by their ability to predict, explain, and systematize what is directly observable. Likewise, Russell claims, mathematical axioms can be justified by their ability to entail, explain, and systematize more obvious mathematical propositions.

Gödel writes approvingly of Russell's analogy between pure mathematics and the empirical sciences.

> [Russell] compares the axioms of logic and mathematics with the laws of nature and logical evidence with sense perception, so that the axioms need not necessarily be evident in themselves, but rather their justification lies (exactly as in physics) in the fact that they make it possible for these "sense perceptions" to be deduced; which of course would not exclude that they also have a kind of intrinsic plausibility similar to that in physics. (1944, p. 449)

The view is summarized by a simple analogy. *Mathematical axioms stand to mathematical data as the laws of nature stand to sensory observations.* The analogy extends to ontology as well.

Gödel compares the relation between mathematical objects and the mathematical data with the relation between physical bodies and sensory observations: both kinds of object are required to obtain a satisfactory theory of the relevant form of "data."[3]

As should be clear, Russell and Gödel operate with a two-tiered conception of evidence. In the empirical sciences, there is *direct evidence* in the form of sense perception, while in mathematics, such evidence takes the form of mathematical intuition or what we are calling "intrinsic" evidence. Then there is *indirect evidence*, which need not be obvious or compelling, or, more generally, enjoy whatever privileged status is enjoyed by the direct evidence. In mathematics, indirect evidence takes the form of what we are calling "extrinsic" evidence. An argument based on extrinsic evidence is thus much like an inference to the best explanation. We have a certain pool of mathematical data, and we give credence to those hypotheses that best predict, explain, and systematize the data.

Gödel's work offers examples of several different forms of extrinsic evidence. One form has to do with a proposition's having plausible elementary consequences, say in the field of Diophantine equations (that is, polynomial equations with integer solutions). As Gödel mentions, some large cardinal axioms enjoy this sort of evidence. For example, the claim that there are "strongly inaccessible" cardinals is far richer in elementary consequences than its negation. Another form of extrinsic evidence accrues to propositions that have abundant "verifiable" consequences and perhaps also enable us to simplify their proofs. A third form has to do with systematicity. In mathematics as much as in physics, there is a need for hypotheses to systematize and explain the data. The laws of Newtonian physics systematize and explain Kepler's laws, for example, and the same goes for certain set-theoretic principles. Finally, there may be extrinsic evidence resulting from the use of broadly inductive methods, involving the extrapolation from examined instances of a general claim.[4]

---

[3] See Gödel (1944, p. 460; 1964, pp. 484–85).

[4] See Gödel (1964, pp. 483, 477, 477 again, and 1951, p. 313, respectively).

12.4 THE THREAT OF PLURALISM

Which forms of appeal to extrinsic evidence are permissible? Might such appeals rest on substantive presuppositions, which aren't always satisfied? We shall now describe a pluralist conception of mathematics and see how such a conception undermines the legitimacy of certain forms of appeal to extrinsic evidence.

The clearest example of pluralism in mathematics comes from geometry, which used to be regarded as the mathematical study of *physical* space, that is, the space that we inhabit. This view became untenable as a result of the development of non-Euclidean geometries in the 1830s (cf. §3.4). There may well be a single true theory of physical space, but this is for physicists to tell. The task of mathematical geometers is to investigate the many kinds of *mathematical* space that are possible. Different geometrical theories are not in conflict with one another but are true of different spaces. Thus, it makes no sense to ask, without further qualification, whether the parallel postulate is true or false: it is true of some spaces and false of others.

The pluralist conception of geometry makes it illegitimate to appeal to extrinsic evidence in at attempt to determine the properties of a single "true" space. A pluralist can happily accept that some spaces have properties that set them apart from others, such as being mathematically simpler, more natural, or more interesting. For example, the parallel postulate yields a geometry that is indisputably simpler and arguably more natural. But this does not impugn non-Euclidean geometries, which exist "side by side" with Euclidean geometry. What is needed, in connection with the parallel postulate, is not more factual knowledge but a semantic decision: which structure are we talking about? The historical evidence suggests that no such semantic decision has been made. When the link between geometrical discourse and physical space was severed, a variety of geometrical spaces came to be regarded as mathematically on a par.

We can distinguish between a monist and a pluralist conception of set theory as well. Monists hold that the structure of the cumulative hierarchy has been singled out uniquely up to isomorphism. Pluralists deny this. The fact that CH is independent of ZFC is often thought to support set-theoretic pluralism. Gödel provides an interesting example. In his contribution to the proof of the mentioned fact, Gödel defines a model of a particularly well-behaved "constructible" hierarchy, L, within the entire cumulative hierarchy, V. Initially, he appears to have thought that this model supported a pluralist conception of set theory:

> [T]he consistency of the proposition $[V = L]$ (that every set is constructible) is also of interest in its own right, especially because it is very plausible that with $[V = L]$ one is dealing with an absolutely undecidable proposition, on which set theory bifurcates into two different systems, similar to Euclidean and non-Euclidean geometry. (1939, p. 155)

Today, set-theoretic pluralism often takes the form of a "pluriverse" view, which holds that there are many universes of sets, each compatible with our full conception of sets.[5] CH is true in some of these universes and false in others. On this view, CH has a truth-value only relative to some particular universe, and the search for a more absolute answer to CH is just as pointless as the analogous quest concerning the parallel postulate.

Pluralists can point to certain technical facts as relevant to their view. First, there are results that parallel the famous nineteenth-century construction of models of non-Euclidean geometry within a given model of Euclidean geometry. As mentioned, Gödel shows how to construct a model of L within V. A wealth of results are now known about how various "setlike" structures can be realized within one another. Second, according to the completeness theorem (proved in Gödel's doctoral dissertation of 1929), every consistent theory has a model. And much is known about the consistency of various systems of set theory relative to one another. We know, for example, that CH is consistent with ZFC just in case its negation is. These results ensure there is a huge variety of structures of broadly set-theoretic character in terms of which the language of set theory *could* be interpreted. Which of these possible interpretations are

---

[5] See Hamkins (2015) for an accessible presentation.

good interpretations, however, is a separate question. While the technical results raise the question of pluralism, they don't on their own answer it.

Despite his initial flirtation with set-theoretic pluralism, Gödel later repudiates the view. He claims instead that "the situation in set theory is very different from that in geometry" (1964, p. 483). The reason is simple. The pluralist conception of geometry is a result of our discovery that the original interpretation of geometry in terms of physical space is untenable. This discovery forced us to abandon what used to be the intended interpretation of the language of geometry. By contrast, the language of set theory has an intended interpretation in terms of the iterative conception. And no reason has ever arisen to abandon this interpretation. Gödel therefore concludes that "the set-theoretical concepts and theorems describe some well-determined reality, in which Cantor's conjecture [CH] must be either true or false" (ibid., p. 476).

Is Gödel right that the iterative conception specifies a "well-determined reality"? As a warm-up case, consider the question of pluralism about arithmetic. Here too there are technical facts analogous those mentioned in connection with set theory. For one thing, a wealth of countable structures can be modeled inside that of the natural numbers. For another, if Dedekind-Peano arithmetic (PA) is consistent, so is the theory that results from adding the negation of this consistency claim. Hence by the completeness theorem, if the former theory has models, so does the latter. These results mean there is a huge variety of structures in terms of which the language of arithmetic *could* be interpreted. So again, the question of pluralism arises. Have we singled out one of these candidate interpretations, at least up to isomorphism? In the case of arithmetic, there are good reasons to be hopeful that we have. For arguably, our conception of the natural numbers suffices to winnow down the candidates to a single isomorphism class. Consider, for example, the structure that models PA and the negation of the claim that PA is consistent. This structure can be ruled out as unintended as soon as we add to our theory the resources to talk about sets of natural numbers

(cf. §12.2).

The case of set theory is harder. If the iterative conception specifies a "well-determined reality," then this specification provides no practical guidance for our attempt to settle CH (cf. §12.2). Our best hope is that the conception determines an answer, albeit one that is not straightforwardly accessible to us.

12.5 THE QUESTION OF QUASI-CATEGORICITY

This hope seems to be fulfilled thanks to Zermelo's *quasi-categoricity theorem* of 1930, which concerns a second-order version of ZF called *ZF2* (of which more shortly). Given any two models of this theory, Zermelo proves that one is isomorphic to the other or to one of its initial segments. It follows that the only difference between two models concerns the "height" of their respective hierarchies, not the structure of their initial segments. This result is important in connection with CH. As we have seen, CH concerns the size of the powerset of the natural numbers relative to its subsets. Its truth is therefore determined by $V_{\omega+2}$, which contains all of the mentioned sets. It follows that all models of ZF2 settle CH in the same way.

The key to Zermelo's theorem is that his theory is *second-order*: it quantifies not only over sets but also over classes thereof. Moreover, these second-order quantifiers are interpreted *standardly*, that is, as ranging over absolutely all classes from the given domain of sets.[6] In ZF2, we can thus express Separation and Replacement as single axioms rather than axiom schemes. Consider Separation:

(Sep 2) $\forall x \forall Y \exists z \forall u (u \in z \leftrightarrow u \in x \land u \in Y)$

This ensures that the intersection of any set x with any class Y is in turn a set. Sep 2 thus captures the maximality idea that at every stage, we form as many sets as we can.[7]

---

[6] This is how the result evades the Löwenheim-Skolem theorem, which states that any first-order theory with an infinite model has models of any infinite cardinality.

[7] The proof of Zermelo's quasi-categoricity theorem is based on some fairly intuitive ideas. Consider two models M and N of ZF2. Assume that these

While Zermelo's theorem is an established mathematical fact, its philosophical significance is controversial. By relying on the notion of a standard model, the theorem helps itself to one of the most problematic notions in set theory, namely that of an arbitrary subclass of some infinite class of objects. As the intractability of CH demonstrates, this notion gives rise to problems that stubbornly resist all of our onslaughts. In light of this, are we really so sure that we have the requisite grasp on the notion of an arbitrary subclass? Or might the notion be "inherently vague," as Feferman contends? These concerns show that the theorem's apparent support for set-theoretic monism might be undermined by its reliance on a notion whose clarity pluralists can reasonably dispute.

It is therefore interesting to ask whether the appeal to quasi-categoricity might be avoided. Might one give an alternative defense of the belief that the iterative conception describes a "well-determined reality"? Gödel appears to have thought so, writing that "[s]uch a belief is by no means chimerical, since it is possible to point out ways in which the decision of a question, which is undecidable from the usual axioms, might nevertheless be obtained" (1964, p. 476). This suggests that a successful search for new axioms, based in part on extrinsic evidence and abductive

---

have been shown to be isomorphic up to some level α, such that we have an isomorphism $f_\alpha : M_\alpha \to N_\alpha$ of their αth initial segments. If either model is exhausted by this initial segment, we are done. So let us assume that both models have height greater than α. We now come to the heart of the proof. We wish to extend f_α to an isomorphism $f_{\alpha+1}$ of the two extended segments, $M_{\alpha+1}$ and $N_{\alpha+1}$. It turns out there is a natural and unique way to do so. Consider a member x of either extended segment, say $M_{\alpha+1}$. If x was present already in the earlier segment, M_α, we let $f_{\alpha+1}(x)$ be $f_\alpha(x)$. If x was not present in M_α, then at least its elements (according to M) were present there. So f_α correlates these "elements" with some members of the other earlier segment, N_α. Since N_α forms a set in $N_{\alpha+1}$, the maximality idea enshrined in Sep 2 ensures that the mentioned members of N_α too form a set y in $N_{\alpha+1}$. So we let $f_{\alpha+1}$ map x to y. It is straightforward to verify that this mapping is one-to-one and onto, and an isomorphism. Proceeding in this way, we construct isomorphisms of ever larger initial segments of the two models. At limit stages, we let the extended isomorphism be the union of all the previous. We continue in this way until we reach the end of one or both models, which proves the theorem.

reasoning, might bolster the case for set-theoretic monism. But such an argument would be question begging. Just as in the case of geometry, pluralism about set theory would, if true, make it illegitimate to appeal to extrinsic evidence in at attempt to determine the properties of a single "true" universe of sets (cf. §12.4).

A monist might counter that it is *always* good scientific methodology to rely on extrinsic evidence and inference to the best explanation more generally. This methodology needs no stamp of approval from philosophy, say in the form of an assurance that a "well-determined reality" has been specified. In this spirit, Maddy writes that "we needn't concern ourselves with whether or not the CH has a determinate truth-value. ... Instead, we need to assess the prospects of finding a new axiom that is well-suited to the goals of set theory and also settles CH" (Feferman et al., 2000, p. 416). I concede that *at the beginning of any inquiry* it is permissible to assume that a "well-determined reality" has been specified. But this assumption can be undermined *at a later stage* by considerations that arise within science, broadly construed. This is precisely what pluralists contend has happened. Using mathematics, we have discovered a range of possible interpretations of our set-theoretic language. Pluralists contend that we are unable to give a scientifically acceptable account of how our set-theoretic thoughts and practices suffice to winnow down these possible interpretations to a single isomorphism class of acceptable interpretations.

We thus face a dilemma. *Either* we must refute the pluralists' contention. We may, for example, defend the controversial assumptions on which the quasi-categoricity argument relies. This would establish monism and thus also legitimize appeals to extrinsic evidence that presuppose monism. *Or else* we must accept pluralism and be wary of the mentioned appeals to extrinsic evidence. This is not the place to try to resolve the dilemma.

Instead, I wish to end by returning to the question of what is at stake. Suppose a community adopts a new set-theoretic axiom through what *appears* to be an appeal to extrinsic evidence. How the community's mathematical development should be interpreted, however, is not obvious. There are two options. One is that the community did in fact rely on extrinsic evidence about a "true" universe of sets and thus extended their knowledge of this universe. This option presupposes monism. But there is an alternative interpretation, which is open to pluralists as well. Perhaps the community merely refined their conception of sets as a result of discovering that one extension of their previous conception has particularly attractive mathematical properties. In short, the observed mathematical practice seems to permit two different interpretations. On the first interpretation, the community's conception of sets stays fixed, while their knowledge expands. On the second interpretation, it is the conception that expands, while the (nonsemantic) knowledge stays fixed. So long as both interpretations are available, mathematical practice can proceed unaffected by the question of whether monism or pluralism is right.

These reflections suggest that the question of pluralism matters less to mathematical practice than one might initially have thought. After all, monists and pluralists can agree that there is a plethora of mathematical structures that are worth exploring, including many of a broadly set-theoretic character.[8] Which of these structures correspond to the "real" sets may not be so important.

SELECTED FURTHER READING

Gödel (1944, 1964) provides two essential texts. Other excellent texts on the question of new set-theoretic axioms are Maddy (1990) and Koellner (2006) and the debate between some leading philosophers and set-theorists is presented in Feferman et al. (2000). Martin (2001) develops a version of Zermelo's quasi-categoricity theorem and defends its philosophical importance. Hamkins (2015) gives a short and accessible presentation of the "multiverse" view of set theory.

---

[8] Perhaps all that matters concerning a candidate axiom is that it should increase the range of mathematical structures that are available. In technical parlance, the axiom should increase the "interpretability strength" of theories. This aim is greatly facilitated by the fact that all "natural" theories are linearly ordered by interpretability strength. See Steel (2014) for a fascinating, though technically challenging, discussion and defense of this point of view.

Concluding Remarks

WE HAVE COVERED A LOT OF GROUND. A huge variety of philosophical approaches to mathematics have been discussed. While I have always attempted to be fair to the positions under discussion, I have (as mentioned in the introduction) made no attempt to hide my own views. It might be useful to summarize some of the main lessons that have emerged, as well as themes that I have emphasized and that serve to distinguish this introduction to the philosophy of mathematics from others.

Let us begin with the fundamental question of what mathematics is about. Five characteristics have appeared throughout the book. Their importance should be widely recognized.

Abstraction. Mathematics is concerned with abstract features of actual or possible objects or systems of objects. Several different forms of abstraction have been discussed. *Fregean abstraction* (cf. chaps. 2 and 9) is nicely illustrated by Frege's example of directions, where we talk about the abstract feature that two lines share just in case they are parallel. Although it is controversial whether Fregean abstraction gives us access to independently existing abstract objects, I have emphasized that this form of abstraction indisputably defines a permissible and useful way to talk *as if* there are abstract objects such as directions.

Next, a form of abstraction is involved in our talk about *quantities.* When we say that one object has mass 2 kilograms and another has charge 2 coulombs, we are using our highly abstract system of real numbers to express claims about how massive and charged our two objects are relative to certain other objects chosen to serve as units of the two quantities (cf. §7.3). Whether such things as numbers "really exist" is controversial, but there can be no doubt that our use of abstract representations of quantities has been a huge success and enabled a vast amount of good science.

Finally, *structuralism* provides a systematic way to talk and theorize about abstract features of entire systems of objects, as opposed to individual things (cf. §§3.4–3.5 and chap. 11). This is the form of abstraction that is most important in contemporary mathematical practice. The metaphysical significance of this form of abstraction too is disputed. Whereas the abstract manner of speaking and theorizing is well defined (as demonstrated by set-theoretic structuralism), the existence of "pure" structures whose members have no individual natures but are merely "positions" in these pure structures remains controversial.

Idealization. As Plato emphasized, geometry is not concerned with our imperfect drawings of circles but with perfect circles, each point of which is *exactly* the same distance from its center. Likewise, the real number line is assumed to be infinitely divisible and complete, regardless of whether the physical world contains any such lines (something Hilbert doubted, as we saw in §4.3). Huge idealizations are also involved in the assumptions made by the representation theorems that underlie our talk about quantities (cf. §7.3).

Computation. By 'computation,' we understand algorithmic operations on syntactic signs or other systems of representations. As we have seen, computation plays an essential role in connection with term formalism, Hilbert's finitism, and intuitionism (cf. §§3.3, 4.4, and 5.4–5.5, respectively). The mathematical and philosophical importance of computation is indisputable. The signs on which we compute can be used to represent objects and their properties, which makes computation a tool of tremendous power—as has become abundantly clear far outside of academia. By contrast, whether our computations on numerals should be taken to constitute genuine reference to numbers remains controversial.

Extrapolation and infinity. For every numeral, we know how to construct its successor (say, by appending one more stroke to a sequence of Hilbert strokes). So we extrapolate and start to reason about the entire sequence of numerals (or perhaps even the entire sequence of natural numbers). Infinity is thus

184

introduced into the heart of mathematics. At first, the infinities in question may be understood as merely potential. But as Cantor convincingly argued, it is conceptually coherent and mathematically fruitful to go further and countenance actual infinities as well (cf. §4.2). This enables far greater extrapolations, as illustrated by the iterative conception of sets, where the powerset operation is iterated transfinitely many times so as to "form" ever more sets (cf. chap. 10).

Proof. Throughout the book, I have emphasized how proof is primarily *an instrument* in our mathematical investigations, not *the object* of these investigations. Mathematical language has some form of meaning, which is typically bestowed on it by one of the aforementioned characteristics. (Of course, the exact analysis of this meaning is controversial.) It is on the basis of this meaning that we are able to recognize an attempted proof as successful or not. It is only quite late in the history of mathematics that formal systems were formulated and proofs could thus become an object of (contentful) metamathematical investigation (cf. §§2.1–2.2).

I believe all of the above characteristics are important in our answer to the question of what mathematics is about. A philosophy of mathematics that excludes some of the characteristics would be one-sided and incomplete.

As adumbrated in the introduction, the book has some main themes, where some of my own views and preferences become apparent. Since these themes weave through the entire text in a way that I hope is fairly unobtrusive, an explicit summary may be useful.

Frege. Like Frege, I have emphasized that mathematics is an autonomous science. It is *autonomous* because it should not be subsumed under, or unduly assimilated to, the paradigmatic empirical sciences. This involves a rejection of empiricism about mathematics (cf. chap. 6 but also §12.3). It is *a science* because its statements are meaningful and (at least in elementary mathematics) have objective truth-values which are often known by us. This involves a rejection of formalism and of all forms of antirealism that take mathematical truths to depend on our proofs or constructions (cf. chaps. 3 and 5, respectively). I have also followed Frege in emphasizing the mathematical and philosophical importance of abstraction.

Object realism vs. platonism. While an object realist holds that there are mathematical objects such as numbers and sets, a platonist goes further by making the (not very precise) claim that these objects are just as "real" as physical objects (cf. §2.5). Throughout the book, we have encountered various ways in which one can be an object realist without being a full-fledged platonist. One option has already been dismissed, namely that mathematical objects are mind-dependent in a way that sets them apart from physical objects. But there are more promising options. According to Frege, the objectivity of mathematical statements is explanatorily prior to the existence of mathematical objects. While the existence of planets contributes to the explanation of the objectivity of our discourse about planets, Frege took the reverse explanatory order to be more appropriate in mathematics (cf. §2.5). Next, perhaps mathematical objects exist only in a potential manner, which contrasts with the actual mode of existence of ordinary physical objects. This idea is at the heart of the ancient notion of potential infinity (cf. §§4.2, 4.4, 5.4–5.5). A version of the idea survives the post-Cantorian embrace of actual infinities, namely as the claim that the cumulative hierarchy of sets is incompletable or merely potential (cf. §§10.5–10.6). Last, noneliminative structuralists aspire to a conception of mathematical objects as mere positions in structures, in contrast to ordinary physical objects, which exist and have many of their properties independently of the structures to which they belong (cf. §11.3). For example, the author of this book is not tied to his departmental structure as intimately as the number 2 is tied to the natural number structure.

Epistemology subject to the integration challenge. It is not an accident that so many of our mathematical beliefs are true. How, then, are our ways of forming mathematical beliefs connected with their subject matter (cf. §1.5)? We have explored a variety of possible forms of mathematical evidence on the basis of which we form our mathematical beliefs: a quasi-perceptual form of intuition, different forms of conceptual evidence, as well as Gödel's famous notion of extrinsic evidence (cf. chap. 8; chaps. 2, 9, §12.2; and §12.3, respectively). My own orientation has been pluralist and gradualist. There appear to be several different sources of mathematical evidence, which gradually become less secure as they take us into the higher reaches of the subject.

Bibliography

Awodey, S. (2014). Structuralism, invariance, and univalence. *Philosophia Mathematica*, 22(1):1–11.

Baker, A. (2005). Are there genuine mathematical explanations of physical phenomena? *Mind*, 114(454):223–38.

Baker, A. (2015). Non-deductive methods in mathematics. In Zalta, E. N., editor, *The Stanford Encyclopedia of Philosophy*. Fall 2015 edition.

Barrow, J. (2010). Simple really: From simplicity to complexity—and back again. In Bryson, B., editor, *Seeing Further: The Story of Science and the Royal Society*, pages 361–84. Harper, London.

Beaney, M. (1997). *The Frege Reader*. Blackwell, Oxford.

Benacerraf, P. (1965). What numbers could not be. *Philosophical Review*, 74:47–73. Reprinted in Benacerraf and Putnam (1983).

Benacerraf, P. (1973). Mathematical Truth. *Journal of Philosophy*, 70(19):661–79. Reprinted in Benacerraf and Putnam (1983).

Benacerraf, P. and Putnam, H., editors (1983). *Philosophy of Mathematics: Selected Readings*. Cambridge University Press, Cambridge. Second edition.

Bernays, P. (1935). On platonism in mathematics. Reprinted in Benacerraf and Putnam (1983).

Boolos, G. (1971). The iterative conception of set. *Journal of Philosophy*, 68:215–32. Reprinted in Benacerraf and Putnam (1983).

Boolos, G. (1989). Iteration again. *Philosophical Topics*, 17:5–21.

Boolos, G. (1998). Gottlob Frege and the foundations of arithmetic. In *Logic, Logic, and Logic*. Harvard University Press, Cambridge, MA.

Boolos, G., Burgess, J. P., and Jeffrey, C. (2007). *Computability and Logic*. Cambridge University Press, Cambridge.

Bourbaki, N. (1996). The architecture of mathematics. In Ewald, W., editor, *From Kant to Hilbert: A Source Book in the Foundations of Mathematics*, volume 2, pages 1265–76. Oxford University Press, Oxford.

Bibliography

Brouwer, L. (1913). Intuitionism and formalism. *Bulletin of the American Mathematical Society*, 20:81–96. Reprinted in Benacerraf and Putnam (1983).

Brouwer, L. (1928). Intuitionist reflections on formalism. Reprinted in Mancosu (1998).

Brouwer, L. (1949). Consciousness, philosophy, and mathematics. Reprinted in Benacerraf and Putnam (1983).

Burgess, J. P. (1984). Synthetic mechanics. *Journal of Philosophical Logic*, 13(4):379–95.

Burgess, J. P. (1999). Review of Stewart Shapiro, *Philosophy of Mathematics: Structure and Ontology*. *Notre Dame Journal of Formal Logic*, 40(2):283–91.

Burgess, J. P. (2005). *Fixing Frege*. Princeton University Press, Princeton, NJ.

Burgess, J. P. (2015). *Rigor and Structure*. Oxford University Press, Oxford.

Burgess, J. P., and Rosen, G. (1997). *A Subject with No Object*. Oxford University Press, Oxford.

Colyvan, M. (2010). There is no easy road to nominalism. *Mind*, 119(474):285–306.

Colyvan, M. (2015). Indispensability arguments in the philosophy of mathematics. In Zalta, E. N., editor, *The Stanford Encyclopedia of Philosophy*. Spring 2015 edition.

Curry, H. B. (1954). Remarks on the definition and nature of mathematics. *Dialectica*, 8:228–33. Reprinted in Benacerraf and Putnam (1983).

Dedekind, R. (1888). *Was Sind und Was Sollen die Zahlen?* Vieweg, Braunschweig. English translation in Ewald (1996).

Dedekind, R. (1963). Continuity and irrational numbers. In *Essays on the Theory of Numbers*. Dover, New York. First published in 1872 as "Stätigkeit und Irrationale Zahlen."

Dummett, M. (1978a). The philosophical basis of intuitionistic logic. In *Truth and Other Enigmas*, pages 215–47. Harvard University Press, Cambridge, MA. Reprinted in Benacerraf and Putnam (1983).

Dummett, M. (1978b). *Truth and Other Enigmas*. Harvard University Press, Cambridge, MA.

Dummett, M. (1981). *Frege: Philosophy of Language*. Harvard University Press, Cambridge, MA. Second edition.

Dummett, M. (1991). *Frege: Philosophy of Mathematics*. Harvard University Press, Cambridge, MA.

Ewald, W. (1996). *From Kant to Hilbert: A Source Book in the Foundations of Mathematics*, volume 2. Oxford University Press, Oxford.

Feferman, S. (1988). Hilbert's program relativized: Proof-theoretical and foundational reductions. *Journal of Symbolic Logic*, 53(2):364–84.

Feferman, S. (1993). Why a little bit goes a long way: Logical foundations of scientifically applicable mathematics. *Philosophy of Science*, 2:442–55.

Feferman, S. (1999). Does mathematics need new axioms? *American Mathematical Monthly*, 106:99–111.

Feferman, S. (2009). Conceptions of the continuum. *Intellectica*, 51:169–89.

Feferman, S. (2014). The Continuum Hypothesis is neither a definite mathematical problem nor a definite logical problem. Available from his webpage.

Feferman, S., Friedman, H. M., Maddy, P., and Steel, J. R. (2000). Does mathematics need new axioms? *Bulletin of Symbolic Logic*, 6(4):401–46.

Field, H. (1980). *Science without Numbers: A Defense of Nominalism*. Princeton University Press, Princeton, NJ.

Field, H. (1982). Realism and anti-realism about mathematics. *Philosophical Topics*, 13(1):45–69.

Field, H. (1984). Platonism for cheap? Crispin Wright on Frege's Context Principle. *Canadian Journal of Philosophy*, 14:637–62. Reprinted in Field (1989).

Field, H. (1985). On conservatives and incompleteness. *Journal of Philosophy*, 82(5):239–60.

Field, H. (1989). *Realism, Mathematics, and Modality*. Blackwell, Oxford.

Field, H. (1991). Metalogic and modality. *Philosophical Studies*, 62(1):1–22.

Føllesdal, D. (1995). Gödel and Husserl. In Hintikka, J., editor, *From Dedekind to Gödel: Essays on the Development of the Foundations of Mathematics*. Kluwer, Dordrecht.

Frege, G. (1879). Begriffsschrift: Eine der arithmetischen nachgebildete Formelsprache des reinen Denkens. Translated in van Heijenoort (1967).

Frege, G. (1894). Review of Husserl, *Philosophie der Arithmetik*. In Geach, P. and Black, M. editors, *Translations from the Philosophical Works of Gottlob Frege*, Blackwell, Oxford, 1952.

Bibliography

Frege, G. (1953). *Foundations of Arithmetic*. Blackwell, Oxford. Translated by J. L. Austin.

Frege, G. (2013). *Basic Laws of Arithmetic*. Oxford University Press, Oxford. Translated by Philip A. Ebert and Marcus Rossberg.

Friedman, M. (1985). Kant's theory of geometry. *Philosophical Review*, 94(4):455–506.

Friedman, M. (1997). Philosophical naturalism. *Proceedings and Addresses of the American Philosophical Association*, 71(2):5–21.

Gettier, E. (1963). Is justified true belief knowledge? *Analysis*, 23(6): 121–23.

Gödel, K. (1933). The present situation in the foundations of mathematics. In Gödel (1995).

Gödel, K. (1939). Lecture at Göttingen. In Gödel (1995).

Gödel, K. (1944). Russell's Mathematical Logic. In Benacerraf and Putnam (1983).

Gödel, K. (1951). Some basic theorems on the foundations of mathematics and their implications. In Gödel (1995).

Gödel, K. (1964). What is Cantor's Continuum Hypothesis? In Benacerraf and Putnam (1983).

Gödel, K. (1995). *Collected Works*, volume 3. Oxford University Press, Oxford.

Goldman, A. (1967). A causal theory of knowing. *Journal of Philosophy*, 64:355–72.

Hale, B. (1997). Realism and its oppositions. In Hale, B. and Wright, C., editors, *A Companion to the Philosophy of Language*, pages 271–308. Blackwell, Malden, MA.

Hale, B. (2000). Reals by abstraction. *Philosophia Mathematica*, 8(2):100–123. Reprinted in Hale and Wright (2001a).

Hale, B., and Wright, C. (2000). Implicit definition and the a priori. In Boghossian, P., and Peacocke, C., editors, *New Essays on the A Priori*. Oxford University Press, Oxford. Reprinted in Hale and Wright (2001a).

Hale, B., and Wright, C. (2001a). *Reason's Proper Study*. Clarendon, Oxford.

Hale, B., and Wright, C. (2001b). To bury Caesar ... In Hale and Wright (2001a).

Hale, B., and Wright, C. (2009). The metaontology of abstraction. In Chalmers, D., Manley, D., and Wasserman, R., editors,

Metametaphysics: New Essays on the Foundations of Ontology, pages 178–212. Oxford University Press, Oxford.

Hamkins, J. D. (2015). Is the dream solution of the continuum hypothesis attainable? *Notre Dame Journal of Formal Logic*, 56(1):135–45.

Heck, R. (1999). Frege's theorem: An introduction. *Harvard Review of Philosophy*, 7(1):56–73.

Hellman, G. (1989). *Mathematics without Numbers*. Clarendon, Oxford.

Hellman, G. (2001). Three varieties of mathematical structuralism. *Philosophia Mathematica*, 9(3):184–211.

Hersh, R. (1997). *What Is Mathematics, Really?* Oxford University Press, Oxford.

Heyting, A. (1931). Die intuitionistische Grundlegung der Mathematik. *Erkenntnis*, 2:106–15. Translated in Benacerraf and Putnam (1983).

Heyting, A. (1956). Disputation. Reprinted in Benacerraf and Putnam (1983).

Hilbert, D. (1899). *The Foundations of Geometry*. Open Court, Chicago. (This edition published 1921.)

Hilbert, D. (1926). Über das Unendliche. *Mathematische Annalen*, 95:161–90. Translated as "On the Infinite" in van Heijenoort (1967).

Hilbert, D. (1927). The foundations of mathematics. Translated in van Heijenoort (1967).

Hilbert, D. (1935). *Gesammelte Abhandlungen*, volume 3. Springer, Berlin.

Horsten, L. (2016). Philosophy of mathematics. In Zalta, E. N., editor, *The Stanford Encyclopedia of Philosophy*. Summer 2016 edition.

Iemhoff, R. (2015). Intuitionism in the philosophy of mathematics. In Zalta, E. N., editor, *The Stanford Encyclopedia of Philosophy*. Spring 2015 edition.

Kant, I. (1997). *Critique of Pure Reason*. Cambridge University Press, Cambridge. Translated by Paul Guyer and Allen Wood.

Koellner, P. (2006). The question of absolute undecidability. *Philosophia Mathematica*, 14(2):153–88.

Koellner, P. (2013). The continuum hypothesis. In Zalta, E. N., editor, *The Stanford Encyclopedia of Philosophy*. Summer 2013 edition.

Krantz, D., Luce, D., Suppes, P., and Tversky, A. (1971). *Foundations of Measurement, Vol. I: Additive and Polynomial Representations*. Academic Press, New York.

Kreisel, G. (1958). Review of Wittgenstein's remarks on the foundations of mathematics. *British Journal for the Philosophy of Science*, 9:135–58.

Lear, J. (1977). Sets and semantics. *Journal of Philosophy*, 74(2):86–102.

Lewis, D. (1986). *On the Plurality of Worlds*. Blackwell, Oxford.

Linnebo, Ø. *Thin Objects: An Abstractionist Account*. Oxford University Press, Oxford. Forthcoming.

Linnebo, Ø. (2006). Epistemological challenges to mathematical platonism. *Philosophical Studies*, 129(3):545–74.

Linnebo, Ø. (2008). Structuralism and the notion of dependence. *Philosophical Quarterly*, 58:59–79.

Linnebo, Ø. (2009). Introduction [to a special issue on the bad company problem]. *Synthese*, 170(3):321–29.

Linnebo, Ø. (2010). Pluralities and sets. *Journal of Philosophy*, 107(3):144–64.

Linnebo, Ø. (2012). Plural quantification. In *Stanford Encyclopedia of Philosophy*, available at http://plato.stanford.edu/archives/fall2012/entries/plural-quant/.

Linnebo, Ø. (2013a). Platonism in the philosophy of mathematics. In Zalta, E. N., editor, *The Stanford Encyclopedia of Philosophy*. Winter 2013 edition.

Linnebo, Ø. (2013b). The potential hierarchy of sets. *Review of Symbolic Logic*, 6(2):205–28.

Linnebo, Ø., and Pettigrew, R. (2014). Two types of abstraction for structuralism. *Philosophical Quarterly*, 64(255):267–83.

Linnebo, Ø., and Rayo, A. (2012). Hierarchies ontological and ideological. *Mind*, 121(482):269–308.

Linnebo, Ø., and Shapiro, S. (2016). Actual and potential infinity. Unpublished manuscript.

MacBride, F. (2003). Speaking with shadows: A study of neo-logicism. *British Journal for the Philosophy of Science*, 54(1):103–63.

Maddy, P. (1990). *Realism in Mathematics*. Clarendon, Oxford.

Maddy, P. (1992). Indispensability and practice. *Journal of Philosophy*, 89(6):275–89.

Maddy, P. (1997). *Naturalism in Mathematics*. Clarendon, Oxford.

Malament, D. (1982). Review of Field's *Science without Numbers*. *Journal of Philosophy*, 79(9):523–34.

Mancosu, P. (1998). *From Brouwer to Hilbert: The Debate on the Foundations of Mathematics in the 1920s*. Oxford University Press, Oxford.

Mancosu, P. (2008). *The Philosophy of Mathematical Practice*. Oxford University Press, Oxford.

Mancosu, P. (2015). Explanation in mathematics. In Zalta, E. N., editor, *The Stanford Encyclopedia of Philosophy*. Summer 2015 edition.

Marquis, J.-P. (2015). Category theory. In Zalta, E. N., editor, *The Stanford Encyclopedia of Philosophy*. Winter 2015 edition.

Martin, D. A. (2001). Multiple universes of sets and indeterminate truth values. *Topoi*, 20(1):5–16.

McCarty, D. C. (2005). Intuitionism in mathematics. In Shapiro, S., editor, *Oxford Handbook of Philosophy of Mathematics and Logic*, pages 356–86. Oxford University Press, Oxford.

McLarty, C. (2004). Exploring categorical structuralism. *Philosophia Mathematica*, 12(1):37–53.

Melia, J. (1995). On what there's not. *Analysis*, 55(4):223–29.

Moschovakis, J. (2015). Intuitionistic logic. In Zalta, E. N., editor, *The Stanford Encyclopedia of Philosophy*. Spring 2015 edition.

Parsons, C. (1977). What is the iterative conception of set? In Butts, R. and Hintikka, J., editors, *Logic, Foundations of Mathematics, and Computability Theory*, pages 335–67. Reidel, Dordrecht. Reprinted in Benacerraf and Putnam (1983) and Parsons (1983).

Parsons, C. (1980). Mathematical intuition. *Proceedings of the Aristotelian Society*, 80:145–68.

Parsons, C. (1982). Kant's philosophy of arithmetic. In Parsons (1983).

Parsons, C. (1983). *Mathematics in Philosophy*. Cornell University Press, Ithaca, NY.

Parsons, C. (1990). The structuralist view of mathematical objects. *Synthese*, 84:303–46.

Parsons, C. (2008). *Mathematical Thought and Its Objects*. Cambridge University Press, Cambridge.

Paseau, A. (2007). Boolos on the justification of set theory. *Philosophia Mathematica*, 15(1):30–53.

Peacocke, C. (1999). *Being Known*. Oxford University Press, Oxford.

Posy, C. (2005). Intuitionism and philosophy. In Shapiro, S., editor, *The Oxford Handbook of Philosophy of Mathematics and Logic*, pages 319–55. Oxford University Press, Oxford.

Potter, M. D. (2004). *Set Theory and Its Philosophy: A Critical Introduction*. Oxford University Press, Oxford.

Putnam, H. (1967a). Mathematics without foundations. *Journal of Philosophy*, 64(1):5–22. Reprinted in Benacerraf and Putnam (1983) and Putnam (1975b).

Putnam, H. (1967b). The thesis that mathematics is logic. In Schoenman, R., editor, *Bertrand Russell, Philosopher of the Century*. Allen and Unwin, London. Reprinted in Putnam (1975b).

Putnam, H. (1971). *Philosophy of Logic*. Harper and Row, New York.

Putnam, H. (1975a). The analytic and synthetic. In *Mind, Language and Reality: Philosophical Papers*, volume 2, pages 33–69. Cambridge University Press, Cambridge.

Putnam, H. (1975b). *Mathematics, Matter and Method*. Cambridge University Press, Cambridge.

Putnam, H. (2012). *Philosophy in the Age of Science: Physics, Mathematics, and Skepticism*. Harvard University Press, Cambridge, MA.

Quine, W. V. (1953a). *From a Logical Point of View*. Harvard University Press, Cambridge, MA.

Quine, W. V. (1953b). Two dogmas of empiricism. In *From a Logical Point of View*. Harvard University Press, Cambridge, MA.

Quine, W. V. (1960). *Word and Object*. MIT Press, Cambridge, MA.

Quine, W. V. (1986). Reply to Charles Parsons. In Hahn, L. E., and Schilpp, P. A., editors, *The Philosophy of W. V. Quine*, pages 396–403. Open Court, Chicago.

Quine, W. V. (1992). *Pursuit of Truth*. Harvard University Press, Cambridge, MA. Second edition.

Quine, W. V. (1995). *From Stimulus to Science*. Harvard University Press, Cambridge, MA.

Raatikainen, P. (2015). Gödel's incompleteness theorems. In Zalta, E. N., editor, *The Stanford Encyclopedia of Philosophy*. Spring 2015 edition.

Ramsey, F. (1931). The foundations of mathematics. In Braithwaite, R., editor, *The Foundations of Mathematics and Other Essays*. Routledge & Kegan Paul, London.

Rayo, A. (2016). Neo-Fregeanism reconsidered. In Ebert, P., and Rossberg, M., editors, *Abstractionism: Essays in Philosophy of Mathematics*: Oxford.

Reck, E. H. (2003). Dedekind's structuralism: An interpretation and partial defense. *Synthese*, 137(3):369–419.

Resnik, M. (1980). *Frege and the Philosophy of Mathematics*. Cornell University Press, Ithaca, NY.

Resnik, M. (1981). Mathematics as a science of patterns: Ontology and reference. *Nous*, 15:529–50.

Rodych, V. (2011). Wittgenstein's philosophy of mathematics. In Zalta, E. N., editor, *The Stanford Encyclopedia of Philosophy*. Summer 2011 edition.

Rosen, G. (2014). Abstract objects. In Zalta, E. N., editor, *The Stanford Encyclopedia of Philosophy*. Fall 2014 edition.

Russell, B. (1903). *Principles of Mathematics*. Norton, New York.

Russell, B. (1907). The regressive method of discovering the premises of mathematics. In Russell, B. (1973). *Essays in Analysis* (edited by Douglas Lackey). George Braziller, New York, pages 272–83.

Russell, B. (1912). *The Problems of Philosophy*. Barnes & Noble Books, New York.

Shapiro, S. (1983). Conservativeness and incompleteness. *Journal of Philosophy*, 80(9):521–31.

Shapiro, S. (1993). Modality and ontology. *Mind*, 102(407):455–81.

Shapiro, S. (1997). *Philosophy of Mathematics: Structure and Ontology*. Oxford University Press, Oxford.

Shapiro, S. (2000). *Thinking about Mathematics*. Oxford University Press, Oxford.

Shapiro, S. (2003). Prolegomenon to any future neo-logicist set theory: Extensionality and indefinite extensibility. *British Journal for the Philosophy of Science*, 54(1):59–91.

Shapiro, S. (2005a). Higher-order logic. In Shapiro, S., editor, *Oxford Handbook of Philosophy of Mathematics and Logic*, pages 751–80. Oxford University Press, Oxford.

Shapiro, S. (2005b). *The Oxford Handbook of Philosophy of Mathematics and Logic*. Oxford University Press, Oxford.

Shapiro, S. (2006). Structure and identity. In MacBride, F., editor, *Identity and Modality*, pages 109–45. Clarendon, Oxford, Oxford.

Sieg, W. (2013). *Hilbert's Programs and Beyond*. Oxford University Press, Oxford.

Skorupski, J. (1989). *John Stuart Mill*. Routledge, London.

Steel, J. R. (2014). Gödel's program. In Kennedy, J., editor, *Interpreting Gödel: Critical Essays*, pages 153–79. Cambridge University Press, Cambridge.

Studd, J. P. (2016). Abstraction reconceived. *British Journal for the Philosophy of Science*, 67(2):579–615.

Bibliography

Tait, W. W. (1981). Finitism. *Journal of Philosophy*, 78(9):524–46.

Tieszen, R. (1989). *Mathematical Intuition: Phenomenology and Mathematical Knowledge*. Kluwer, Dordrecht.

Troelstra, A. (1998). Kleene's realizability. In Buss, S. R., editor, *Handbook of Proof Theory*, pages 407–73. Elsevier, Amsterdam.

van Atten, M. (2004). *On Brouwer*. Wadsworth, Belmont, CA.

van Atten, M. (2014). The development of intuitionistic logic. In Zalta, E. N., editor, *The Stanford Encyclopedia of Philosophy*. Spring 2014 edition.

van Heijenoort, J., editor (1967). *From Frege to Gödel*. Harvard University Press, Cambridge, MA.

von Neumann, J. (1931). The formalist foundations of mathematics. Translated in Benacerraf and Putnam (1983).

Weir, A. (2015). Formalism in the philosophy of mathematics. In Zalta, E. N., editor, *The Stanford Encyclopedia of Philosophy*. Spring 2015 edition.

Weyl, H. (1949). *Philosophy of Mathematics and Natural Science*. Princeton University Press, Princeton, NJ.

Wright, C. (1983). *Frege's Conception of Numbers as Objects*. Aberdeen University Press, Aberdeen.

Wright, C. (1997). The philosophical significance of Frege's theorem. In Heck, R., editor, *Language, Thought, and Logic. Essays in Honour of Michael Dummett*. Clarendon, Oxford. Reprinted in Hale and Wright (2001a).

Yablo, S. (2005). The myth of the seven. In Kalderon, M., editor, *Fictionalism in Metaphysics*, pages 88–115. Oxford University Press, Oxford.

Yablo, S. (2014). *Aboutness*. Princeton University Press, Princeton, NJ.

Zach, R. (2015). Hilbert's program. In Zalta, E. N., editor, *The Stanford Encyclopedia of Philosophy*. Summer 2015 edition.

Zermelo, E. (1908). A new proof of the possibility of a well-ordering. Translated in van Heijenoort (1967).

Zermelo, E. (1930). Über Grenzzahlen und Mengenbereiche. *Fundamenta Mathematicae*, 16:29–47. Translated in Ewald (1996).

Index

Index

Printed and bound by CPI Group (UK) Ltd, Croydon, CR0 4YY

27/10/2024

14580232-0002